D1300661

Springer Tracts in Modern Physics 114

Springer Tracts in Modern Physics

90* **Jets of Hadrons** By W. Hofmann

91 **Structural Studies of Surfaces**
With contributions by K. Heinz, K. Müller, T. Engel, and K. H. Rieder

92 **Single-Particle Rotations in Molecular Crystals** By W. Press

93 **Coherent Inelastic Neutron Scattering in Lattice Dynamics** By B. Dorner

94 **Exciton Dynamics in Molecular Crystals and Aggregates** With contributions by V. M. Kenkre and P. Reineker

95 **Projection Operator Techniques in Nonequilibrium Statistical Mechanics**
By H. Grabert

96 **Hyperfine Structure in 4d- and 5d-Shell Atoms** By S. Büttgenbach

97 **Elements of Flow and Diffusion Processes in Separation Nozzles** By W. Ehrfeld

98 **Narrow-Gap Semiconductors** With contributions by R. Dornhaus, G. Nimtz, and B. Schlicht

99 **Dynamical Properties of IV–VI Compounds** With contributions by H. Bilz, A. Bussmann-Holder, W. Jantsch, and P. Vogl

100* **Quarks and Nuclear Forces** Edited by D. C. Fries and B. Zeitnitz

101 **Neutron Scattering and Muon Spin Rotation** With contributions by R. E. Lechner, D. Richter, and C. Riekel

102 **Theory of Jets in Electron-Positron Annihilation** By G. Kramer

103 **Rare Gas Solids** With contributions by H. Coufal, E. Lüscher, H. Micklitz, and R. E. Norberg

104 **Surface Enhanced Raman Vibrational Studies at Solid/Gas Interfaces** By I. Pockrand

105 **Two-Photon Physics at e^+e^- Storage Rings** By H. Kolanoski

106 **Polarized Electrons at Surfaces** By J. Kirschner

107 **Electronic Excitations in Condensed Rare Gases**
By N. Schwentner, E.-E. Koch, and J. Jortner

108 **Particles and Detectors** Festschrift for Jack Steinberger
Edited by K. Kleinknecht and T. D. Lee

109 **Metal Optics Near the Plasma Frequency**
By F. Forstmann and R. R. Gerhardts

110* **Electrodynamics of the Semiconductor Band Edge**
By A. Stahl and I. Balslev

111 **Surface Plasmons on Smooth and Rough Surfaces and on Gratings** By H. Raether

112 **Tests of the Standard Theory of Electroweak Interactions** By C. Kiesling

113 **Radiative Transfer in Nontransparent, Dispersed Media**
By H. Reiss

114 **Electronic Transport in Hydrogenated Amorphous Semiconductors**
By H. Overhof and P. Thomas

115 **Scattering of Thermal Energy Atoms from Disordered Surfaces**
By B. Poelsema and G. Comsa

---

* denotes a volume which contains a Classified Index starting from Volume 36

H. Overhof P. Thomas

Electronic Transport in Hydrogenated Amorphous Semiconductors

With 65 Figures

Springer-Verlag
Berlin Heidelberg New York
London Paris Tokyo

Professor Dr. Harald Overhof

Fachbereich Physik, Universität – Gesamthochschule Paderborn, Warburger Straße 100
D-4790 Paderborn, Fed. Rep. of Germany

Professor Dr. Peter Thomas

Fachbereich Physik, Philipps-Universität, Renthof 5
D-3550 Marburg, Fed. Rep. of Germany

Manuscripts for publication should be addressed to:

Gerhard Höhler

Institut für Theoretische Kernphysik der Universität Karlsruhe
Postfach 6980, D-7500 Karlsruhe 1, Fed. Rep. of Germany

Proofs and all correspondence concerning papers in the process of publication should be addressed to:

Ernst A. Niekisch

Haubourdinstraße 6, D-5170 Jülich 1, Fed. Rep. of Germany

ISBN 3-540-50186-X Springer-Verlag Berlin Heidelberg New York
ISBN 0-387-50186-X Springer-Verlag New York Berlin Heidelberg

Library of Congress Cataloging-in-Publication Data. Overhof, H. (Harald), 1942–. Electronic transport in hydrogenated amorphous semiconductors. (Springer tracts in modern physics; 114) Bibliography: p. 1. Amorphous semiconductors. 2. Transport theory. 3. Electronic structure. I. Thomas, P. (Peter), 1942–. II. Title. III. Series. QC1.S797 vol. 114 530 s [537.6'22] 88-24866 [QC611.8.A5]

Printing and binding: Brühlsche Universitätsdruckerei, 6300 Giessen
2157/3150-543210

Preface

Systematic research on amorphous semiconductors started more than thirty years ago. At that time the motivation for research was mainly academic interest in a field which was still new to researchers because of the absence of periodicity and the presence of disorder: Deviations of the electronic properties of amorphous semiconductors from those of crystalline semiconductors were observed, and theoretical concepts and models were developed which rested on well-known principles that had been used with success in the field of crystalline semiconductors. On the other hand, new theoretical methods were also introduced which explicitly took care of the disorder.

A dramatic rise in experimental activities started when it was realized that under suitable preparation conditions amorphous semiconductors, in particular hydrogenated amorphous silicon, could be used as a cheap basis material for large-scale electronic thin-film devices. Some new and surprising features of electronic properties, especially transport properties, were found. The theoretical models and methods, however, still remained at a somewhat immature state, which in view of the high complexity of the real amorphous semiconducting materials and the lack of a well-defined microscopic symmetry is not surprising. In the last few years, however, a coherent theoretical picture has begun to emerge which allows the interpretation of equilibrium transport properties of amorphous semiconductors to be placed on a more general footing.

This book is devoted to a detailed presentation of the model, the basic theoretical concepts and the interpretation of equilibrium transport data of tetrahedrally coordinated amorphous semiconductors. Most of the material covered by this book has been published in the last few years and is scattered over the literature. We found it therefore useful to give a coherent presentation of our approach, which also includes the discussion of other theoretical work in this field. The book is organized in such a way as to allow the occasional reader not interested in theoretical details to follow the main ideas, while theoreticians are invited to concentrate on the theoretical chapters and sections.

The authors are indebted to Professor Sir Nevill Mott for numerous discussions concerning the prefactor problem of the activated conductivity, and to Professor W. Götze for helpful discussions about his mode-coupling theory of Anderson localization. We thank our colleagues in Marburg, Professor J. Stuke, Professor W. Fuhs and Dr. H. Mell, as well as Dr. W. Beyer from Jülich, and their coworkers, for their continuous and fruitful cooperation in all problems related to experiments. We have benefitted from many fruitful discussions with Professors

H. Fritzsche, E.N. Economou, W.E. Spear and P.G. LeComber. From discussions with Dr. K. Kempter and his colleagues at Siemens AG we learnt many details related to the applications of amorphous semiconductors. Last but not least we thank our numerous students for their valuable and creative contributions to the present subject over the past decade, and Mrs. Anastasia Leventi for her gentle linguistic assistance.

Paderborn and Marburg,
October 1988

H. Overhof
P. Thomas

Contents

List of Symbols*

a	lattice constant of the site lattice	63
A, A^*	heat of transport term of the thermoelectric power	27
A_{iq}	electron-phonon coupling constant	64
$a_{\rm loc}$	decay length of a localized basis state	29
$B_{\rm MNR}$	intercept of the Meyer–Neldel rule	128
$\boldsymbol{B}$	magnetic (induction) field	
b_q , $b_q^\dagger$	phonon annihilation (creation) operator	64
c_i , $c_i^\dagger$	electron annihilation (creation) operator	16
d	dimensionality of the system	
d_i , $d_i^\dagger$	annihilation (creation) operator for displaced oscillators	123
e	elementary charge	
E	energy	
$E_{\rm C}$, $E'_{\rm C}$	mobility edge in the conduction band	17
$E_{\rm F}$	Fermi energy	
$E_{\rm F}^{\rm q}$	quasi-Fermi energy	51
$E_{\rm g}^{\rm opt}$	optical band gap	11
$E_{\rm H}^*$	activation energy of the Hall effect	50
$E_{\rm m}$	mobility edge	27
$E_{\rm MNR}$	characteristic energy of the Meyer–Neldel rule	41
E_Q , E_Q^*	activation energy of the Q function	46
E_S , E_S^*	activation energy of the thermoelectric power	28
$E_{\rm t}$, $E_{\rm t}^*$	dominant transport energy	76
$E_{\rm T}$	slope of the tail density of states distribution	75
$E_{\rm V}$	mobility edge of the valence band	24
E_σ , E_σ^*	activation energy of the conductivity	26
$\boldsymbol{F}$, $\boldsymbol{F}_0$	electric field	
$f_0^{\rm e}$, $f_0^{\rm h}$	Boltzmann distribution function for electrons (holes)	24
$f_{\rm F}(E, E_{\rm F}, T)$	Fermi distribution function	24
$g(E)$	density of states distribution	13
$g(L)$	dimensionless conductance of a sample of linear size L	19
$g_{\rm C}(E)$	conduction band density of states distribution	13
$g_{\rm V}(E)$	valence band density of states distribution	13
H , H'	Hamiltonian	

* Symbols that are of local importance only are not listed here.

H_e , H_{ep} , H_p	electronic, electron-phonon, and phonon contribution to H	63, 64
$\hbar$	Planck's constant	
I	current	
I_0^* , I_{00}	prefactor of the current	53
I_D	drift current of the time-of-flight experiment	56
I_{FB}	flat band current of the field effect experiment	131
I_{SD}	Source-drain current of the field effect experiment	54
$\boldsymbol{j}$	electric current density	
J , J_{ij}	transfer matrix element	16
k	Boltzmann's constant	
$\boldsymbol{k}$	quantum number for electronic states	17
k_F	Fermi wave number	18
l	electrode distance	51
l_{free}	mean free path	18
L_i	inelastic scattering length	22
L_{ij}	Onsager's transport coefficients	23
l_{max}	maximum fluctuation length of the long-ranged potential	110
l_{min}	minimum mean free path	18
m	mass of the electron	
n	electron density	51
N_C , N_V	effective density of extended conduction (valence) states	26
n_i	particle number operator	16
$p(V)$	distribution function of random potentials	115
$P(\varepsilon_i)$	distribution function of site energies	63
p_c	critical bond concentration in percolation theory	31
q	charge of a particle ($\pm e$)	25
$\boldsymbol{q}$	quantum number for phonons	16
Q	combination of conductivity and thermoelectric power	28
Q_0 , Q_0^*	intercept of the Q function	46
r	parameter for the electron-phonon coupling strength	74
R_c	critical radius in the percolation analysis for hopping	31
$\boldsymbol{R}_i$	position of site i	16
R_{ij}	distance between sites i and j	30
s	frequency exponent of the ac conductivity	59
S	thermoelectric power, Seebeck coefficient	27
T	temperature	
T_{ac}	characteristic temperature of the ac conductivity data	60
T_{SLC}	characteristic temperature in the superlinear current	119
T_{TOF}	characteristic temperature of the TOF data	58
T_0	slope parameter of variable range hopping formula	33
t	time	
t_T	transit time in time-of-flight experiment	56
U_i , U_{eff}	Hubbard's correlation energy	123
V	voltage, potential	
$\boldsymbol{v}$	velocity	
V_D	applied voltage of the TOF experiment	56

V_G	gate voltage of field effect 54
V_j	external potential 65
V_{SD}	Source-Drain voltage of field effect 54
$\boldsymbol{w}$	heat current density 23
W	width of the distribution of disorder energies 17
y	coupling constant in the localization theory 68
z	complex frequency

α_1, α_2	dispersion parameters of the TOF data 56
β	$1/kT$
$\beta(g)$	scaling function 19
γ_F^*	temperature coefficient of the Fermi energy 127
$\Gamma_{ij}, \Gamma_{ij}^0$	hopping rate 30
γ_t^*	temperature coefficient of the transport energy 76
δ	rms potential fluctuation 111
ΔE_{opt}	slope of the Urbach tail 13
ΔE_σ^*	change of the apparent activation energy of the conductivity 78
ε''	imaginary part of the dielectric constant 13
ε_i	site energy 16
$\varepsilon_s, \varepsilon_d$	static dielectric constant of the semiconductor (dielectric)
ε_0	vacuum dielectric constant
η	rms width of the distribution of site energies 63
θ	dimensionless temperature 74
λ	critical exponent 20
Λ	phonon propagator 74
μ, μ_0	mobility 25
μ_H, μ_{H0}	Hall mobility 49, 116
ξ	correlation length 21
ξ_{loc}	localization length of an Anderson localized state 19
Π	Peltier coefficient 27
$\varrho(E)$	density of electronic states per site 74
σ	conductivity
$\sigma(E)$	differential conductivity 36
$\sigma_{ext}, \sigma_{hopp}$	extended states (hopping) contribution to σ 38
σ_M	minimum metallic conductivity 18
σ_h, σ_t	hopping (tunneling) contribution to σ 104
σ_0, σ_0^*	prefactor of the conductivity 26
τ_{deph}	dephasing time 96
τ_j, τ_n	current (density) relaxation time 66
χ, χ_{ij}	generalized susceptibility 65
$\Psi_{\boldsymbol{k}}$	electronic wave function 17
Ω	volume of the system
$\omega_{\boldsymbol{q}}$	phonon frequency 64
ω_0	frequency of order phonon frequency 59

List of Acronyms

CFO	Cohen–Fritzsche–Ovshinsky Model 4
CRN	Continuous Random Network 12
ESR	Electron Spin Resonance 14
fcc	face centered cubic 10
gd	glow discharge 5
MIS	Metal Insulator Semiconductor device 54
MNR	Meyer–Neldel Rule 42
NMR	Nuclear Magnetic Resonance 15
TOF	Time-of-Flight 55
RDF	Radial Distribution Function 12
SAW	Surface Acoustical Wave 146
SCLC	Space Charge Limited Current 51
SIMS	Secondary Ion Mass Spectroscopy 15
SLC	Super Linear Current 118
SWE	Staebler–Wronski Effect 42
TROK	Tietje–Rose–Orenstein–Kastner model 57
VLSI	Very Large Scale Integrated devices 1

1. Introduction

In nature inorganic solids are nearly always found as crystalline or polycrystalline materials. Rare exceptions are obsidian and some other volcanic materials which have an amorphous, non-crystalline atomic structure. An amorphous solid is in a metastable state which requires a sufficiently fast quenching from the melt to avoid crystallization.

Man-made amorphous solids have been used since the days of the Babylonians as ordinary glasses because of their unique optical properties: Oxide glasses are highly transparent and quite insensitive to large impurity concentrations. Furthermore, one can fabricate glasses for large scale optical devices which do not show imperfections like the grain boundaries in polycrystalline materials.

For present day technology crystalline semiconductors are used as basic materials. Silicon, in particular, is often considered to be of comparable impact to modern civilisation as, e.g., iron was some three thousand years ago. Semiconductor devices made from crystalline silicon are omnipresent now and very large scale integration (VLSI) on the basis of crystalline silicon is a leading technology of our days. However, our present standard of technology also imposes limitations to the size in which crystalline semiconductor devices can be produced economically.

In order to display information in a way suitable for direct human perception an interface would be needed of about the same scale as that of, e.g., A4–size paper. Amorphous semiconductors can be prepared in virtually any size and have, therefore, been suggested as a material to bridge the gap between small size crystalline VLSI-devices and the human scale. Another well-known application of amorphous semiconductors are inexpensive large area thin-film solar cells.

Large scale electronic devices often require only moderate material quality. If they are to be fabricated from amorphous semiconductors the question arises whether we are able to combine the good properties of semiconductors with the good properties of glasses. The theory and thus the understanding of the solid state has been restricted to crystalline solids for a long time, because for ordered highly symmetric systems there are powerful theoretical tools available. One might ask what happens to the electronic properties of a solid if its long-range order is relaxed, while the short-range order is preserved to a certain degree. One might suspect that all electronic levels become smeared out so that all structures in the relevant spectra become broadened. In this respect an amorphous semiconductor would be a "dirty" counterpart of the respective crystalline solid and in many respects similar to the latter.

A more detailed investigation of the electronic properties of amorphous semiconductors, however, reveals that these materials show a number of entirely new phenomena which lead to interesting theoretical problems and motivate fundamental research.

1.1 Some Unique Properties of Amorphous Semiconductors

i) From a technological point of view, as a basis material for electronic devices, amorphous semiconductors are orders of magnitude less sensitive to impurities than crystalline semiconductors. Extremely high concentrations of dopants and foreign atoms can, and must, be admitted without destroying reasonable device quality. The fabrication of very large area devices presents no insurmountable problems.

ii) Experiments which probe the decay of a non-equilibrium distribution of carriers into the equilibrium state of the system reveal an extremely wide spectrum of relaxation rates, ranging from the sub-picosecond time scale up to days. Very short times are characteristic for phase relaxation of excitations due to scattering at the static disorder. The relaxation towards states close to the mobility edge is in the ps range. Energy relaxation proceeds further by hopping in localized tail states, with times strongly dependent on the density of states.

iii) The equilibrium transport coefficients: conductivity σ, and Seebeck coefficient S, can be combined in a function $Q = \ln(\sigma\,\Omega\,\mathrm{cm}) + qS/k$, where q and k are the charge and Boltzmann's constant, respectively. It turns out that this function has two unexpected features for amorphous semiconductors. When plotted against $1/T$, its intercept Q_0^* is almost universally constant. In contrast to what one should expect on the basis of the conventional transport model in semiconductors, its slope E_Q^* is finite and sample dependent.

iv) In contrast to the nearly universal Q_0^*, the intercept σ_0^* of an Arrhenius plot ($\ln\sigma$ vs. T^{-1}) of the dc conductivity shows a very wide variation, over at least 5 orders of magnitude. It is strongly correlated with the activation energy E_σ^*. A logarithmic plot of σ_0^* vs. E_σ^* is to a good approximation characterized by a straight line with slope E_{MNR}^{-1}. This slope has a remarkably constant value, close to $25\,\mathrm{eV}^{-1}$, at least for films grown by the glow-discharge technique. This type of correlation is referred to as the Meyer–Neldel rule. Meyer and Neldel [1937] found such a rule studying metal oxide powders.

v) At very low temperatures amorphous semiconductors show fascinating thermodynamic properties [Zeller and Pohl, 1971; v. Löhneysen, 1983], which have all the properties of a distribution of two-level systems. The physical origin of these two-level systems in amorphous semiconductors is, however, not at all clear.

1.2 Historical Notes

The physics of crystalline solids is an old and well-developed field. Great names like v. Laue, Brillouin, Bloch, Wilson, Seitz, Mott, etc. are related to it, to mention just a few. The investigation of electrical and optical properties of amorphous semiconductors, on the other hand, seems to have started only in the early fifties of this century, when the optical spectra of amorphous selenium were compared to those of its crystalline counterpart [Stuke, 1953]. There is, however, a very early report from 1878 on transport properties of crystalline selenium [Sabine, 1878]. In that work "amorphous" selenium is mentioned and considered to be a dielectric. The structure of silicon and germanium films prepared by evaporation was investigated by Hass as early as 1947. Using electron microscopy and electron diffraction he observed that below a certain substrate temperature the structure of the films became amorphous.

Besides amorphous selenium, which in the first part of this century was of technological importance for the fabrication of selenium rectifiers, the properties of amorphous chalcogenide glasses attracted the attention of the Russian School [Goryunova and Kolomiets, 1955]. These materials are now widely used in electrophotography machines. Somewhat later, the switching properties of chalcogenide glasses stimulated considerable research in this field [Ovshinsky, 1968].

The study of amorphous tetrahedrally coordinated semiconductors started with the work of Tauc, Grigorovici and Vancu [1966] on amorphous germanium. In contrast to the amorphous materials mentioned above, which are glasses that can be obtained by quenching from the melt, amorphous germanium and silicon can only be prepared as thin films by the deposition of atomic or molecular species. A number of fundamental questions arose at that time which stimulated experimental and theoretical work. No technical application, however, was obvious since the evaporated samples of those days could not be effectively doped.

In the early days of research on amorphous semiconductors these materials were considered as dirty counterparts of their crystalline modifications. The sharp structures found in the optical spectra of crystalline semiconductors, due to long-range order, are completely smeared out in amorphous semiconductors, even the absorption edges are no longer sharp. A continuum of electronic states resulting from overlapping band tail states was supposed to fill the gap between the valence and the conduction band, which thus became a pseudogap [Cohen et al., 1969]. It was then a puzzling question why, e.g., window glass is transparent. From band structure theory of crystalline semiconductors, physicists used to relate the existence of band gaps to long-range order. How could it then be possible that a solid without long-range order posesses a gap in its electronic spectrum? On the other hand, a chemist would not question the existence of gaps in systems without long-range order. The problem of the existence of gaps has in fact been treated on the basis of a chemical-bond-orbit approach, the Hall–Weaire–Thorpe model [Hall, 1952, 1958; Weaire and Thorpe, 1971]. This chemists point of view proved to be valuable when discussing the spectrum of the valence band states, the form of the band edges, the electronic structure of defects, etc. As an input into these theories a structural model for an amorphous semiconductor was needed. It was generally assumed that there exists an ideal amorphous structure, represented by a defect-free continuous random network [Polk, 1971] of atoms. The existence of such an idealized structure was, however, later questioned [Phillips, 1979].

Although the existence of a wide tailing of the bands into the gap seemed to be proved, the conductivity at elevated temperatures showed an activated behaviour, as is the case in intrinsic crystalline semiconductors with sharp band edges. It became clear that this resulted from the influence of the strong static disorder. Ioffe and Regel [1960] studied charge transport in the strong scattering regime, and Bányai [1964] suggested the application of Anderson localization [Anderson, 1958] to amorphous semiconductors. He postulated the existence of two energies, separating localized from delocalized states in the valence and the conduction bands, respectively. This model of the mobility edges was further developed by Mott [1967] and by Cohen, Fritzsche and Ovshinsky [1969] and was referred to as the Mott-CFO model. For a long time it was the basis for the interpretation of transport data showing activated behaviour (see, e.g., the textbook by Mott and Davis [1971, 1979]).

Another transport regime has been extensively studied both experimentally and theoretically, namely hopping near the Fermi level. In the early amorphous semiconductors prepared by evaporation, a large density of states was present in the pseudogap. These states originated from unsaturated bonds (dangling bonds) and formed an energetic continuum of localized eigenstates of high density. The conductivity at not too high temperatures was found to be taking place among states close to the Fermi level by phonon-assisted transitions (hopping). Mott [1969] suggested the model of Variable Range Hopping to account for the unusual non-activated behaviour of the conductivity, being of the form $\ln \sigma \sim (T_0/T)^{1/(d+1)}$ in d dimensions ($d > 1$). The Variable Range Hopping model has been given a firm basis by applying percolation theory [Ambegaokar et al., 1971] or effective resistor network calculations [Pollak, 1972; Maschke et al., 1974; Butcher and McInnes, 1978; McInnes et al., 1980] in order to perform the configurational average of the solution of the linearized rate equation. Many experimental data, including Hall effect and thermoelectric power, could be more or less satisfactorily explained. For a review see, e.g., the book by Böttger and Bryksin [1985]. On the other hand, there are still some unsolved problems, such as the influence of the intersite Coulomb interaction [Pollak and Efros, 1985] on hopping transport and the clear disagreement between the theoretical and the experimental prefactor of the hopping conductivity. Since theories have been tested successfully against computer simulations this latter problem might be due to the idealized homogeneous models used, while real semiconductors may be inhomogeneous on a mesoscopic scale. This is in particular true for the evaporated samples [Barna et al., 1974], where transport is dominated by hopping processes. Research on evaporated amorphous semiconductors virtually stopped after 1975, and many interesting questions related to Variable Range Hopping have been left open.

At the same time a new era began in the field of amorphous semiconductors. In their pioneering work on the preparation of amorphous silicon films by glow-discharge (gd) decomposition of silane the group at the Standard Telecommunications Laboratories [Sterling and Swann, 1965; Chittick et al., 1969] had shown that amorphous silicon prepared by this method differed in many respects from evaporated amorphous semiconductors: the material showed activated dc conductivity, a good photoconductivity, and its conductivity could be increased by the addition of phosphine to the silane gas prior to decomposition [Chittick et al., 1969]. These superior properties were not attributed to the presence of hydrogen but rather to the fact that the gd decomposition was a much "smoother" deposition technique than evaporation or sputtering techniques. The material was reported to be unstable but even then it is remarkable that besides the Dundee group no other research group has followed this work. In 1975 Spear and LeComber showed that this material could be efficiently doped n-type or p-type and could be prepared as a rather stable material. When the Chicago group [Triska et al., 1975] showed by evolution measurements that this material did indeed contain hydrogen (and is, therefore, referred to as a-Si:H) it was soon recognized that this was the prime reason for the superior properties of the gd-prepared material rather than the gentle decomposition technique.

A dramatic rise in experimental, theoretical and, in particular, technological activities has since occured. Besides the well-known application as a material for inexpensive, large area thin-film solar cells, a wide spectrum of large scale electronic devices made from a-Si:H is currently being developed and even manufactured. Examples are large scale photoconductors as sensor lines for telecopying machines, drums for electrophotography, large area thin-film transistor arrays for liquid crystal displays, and video camera tubes. Fascinating switching properties of a-Si:H films have also been reported recently (see the Proceedings of the recent 12th International Conference on Amorphous and Liquid Semiconductors [Matyáš et al., 1987] for an overview).

In a-Si:H films the density of unsaturated bonds is so low that Variable Range Hopping close to the Fermi level can only be observed at low temperatures in some samples and that the activated behaviour prevails. Looking more carefully at this latter transport regime it has become clear that some features of the transport data cannot be explained using the concept of sharp mobility edges. This is not surprising since Anderson localization simply does not exist in a system with inelastic interactions [Thouless, 1977; Imry, 1980] due to electron-phonon coupling, and at the fairly high temperatures used in order to measure equilibrium transport in a semiconductor. It was, therefore, necessary to study theoretically the transport processes of carriers in systems which are characterized by both strong static disorder and electron-phonon interaction. Two complementary approaches emerged. The first stresses the static aspects of the electron-phonon interaction which can be enhanced by the particular fractal character [Aoki, 1982, 1983; Soukoulis and Economou, 1984] of the electronic eigenstates close to the mobility edge. This approach then predicts the formation of small polarons as the relevant quasiparticles [Anderson, 1972; Cohen et al., 1983, 1984]. The second concentrates on the dynamical aspects of the interaction [Müller and Thomas, 1983, 1984]. Due to the energy transfer from the phonon system to the electron system incoherent hopping processes occur, which in turn lead to a phonon induced delocalization.

These theories are based on models containing static disorder which is only short-range correlated. On the other hand, there is experimental evidence that amorphous semiconductors can possess internal long-range correlated fields. Doped films in particular should contain randomly distributed charges leading to a long-ranged Coulomb field. The existence of such fields should therefore be taken into account when transport or optical data are to be interpreted [Overhof and Beyer, 1981].

At present a coherent physical picture of amorphous tetrahedrally coordinated semiconductors is emerging. This must contain many novel aspects in order to describe the experimental data, which show many features that are unique to the amorphous phase.

1.3 Aim of the Book

The transport properties of amorphous semiconductors, in particular a-Si : H and a-Ge : H, are usually interpreted in terms of phenomenological models which are used to extract the physical parameters that describe the semiconductors. Most researchers, e.g., use the activation energy of the dc conductivity in order to determine the position of the Fermi level with respect to the mobility edges. Such a procedure is allowed if one has a theory that matches the experimental data to such a degree of perfection that the parameters entering the theory can be identified with physical quantities. The very existence of the Meyer–Neldel rule demonstrates that one of the parameters, the prefactor of the conductivity, makes no sense in the phenomenological model. This casts severe doubts on the validity of the other parameter, the activation energy.

In our book we shall use another approach: We start from a transport theory on a microscopic level that treats static disorder and interaction with phonons. This model is then augmented to incorporate the complications that are needed in order to cope with the complexity of real systems. The theoretical concepts and tools are then used to describe a variety of experiments, and it is shown how microscopic parameters can be extracted from experimental data.

It is an essential part of the present work to test in detail our theory by comparing the predictions with a large body of independent experimental data. Therefore, a considerable part of this book will be devoted to the presentation of experimental techniques and results.

1.4 Plan of this Book

The approach chosen for this book does not allow us simply to present the theoretical material by starting from some Hamiltonian and continuing by straightforward calculations. In order to make the book also profitable to experimentalists, we have tried to develop the theory on various levels with increasing completeness and complexity. This will enable us to show that the unique features of amorphous semiconductors cannot be interpreted on the basis of simple physical arguments alone, but require elaborate theoretical approaches.

In Chap. 2 we give a brief introduction to the atomic structure of amorphous tetrahedrally coordinated semiconductors and discuss the present day model for the density of states that has emerged from a large number of experimental information. We introduce the concept of Anderson localization and discuss it in terms of the scaling theory. This leads to a preliminary distinction between transport in the extended and the localized states. The theoretical approaches to the corresponding regimes of transport are discussed in some detail. We define a standard transport model in an ad hoc manner, which is later used as a preliminary basis for the interpretation of the experimental data on activated transport. We also introduce the small polaron model and discuss the approximate description of activated transport processes in terms of the Kubo–Greenwood formula.

Typical experimental results are presented in Chap. 3, including the dc conductivity σ, the Seebeck coefficient S, the Hall mobility μ_{H}, space-charge limited currents, time-of-flight experiments, field effect, and ac conductivity. We point out that the unusual features of these experiments cannot be explained by the standard transport model and require a more refined theory.

We start in Chap. 4 by presenting the basic physics of a transport theory based on a homogeneous model, including strong static disorder and electron-phonon interaction. The theory can be viewed as a generalization of conventional hopping theory. For vanishing electron-phonon coupling and for zero temperature our approach reduces to the mode-coupling theory developed by Götze and coworkers, which describes Anderson localization. On the other hand, for electrons sufficiently below the mobility edge in the conduction band, the theory reduces to the conventional hopping theory in the high-density, high-temperature limit. The enhanced tendency of small polaron formation close to the mobility edge is also discussed. For the homogeneous model treated in this chapter the theory provides a justification for the essential features of the standard transport model.

The details of the theoretical approach and the discussion of the approximations involved are postponed until Chap. 5. The results and their consequences for transport in a semiconductor model are, however, discussed in Chap. 4. The reader not interested in the technical details of the theory may therefore skip Chap. 5 without losing the line of the arguments. The reader more interested in theory will concentrate on Chap. 5, which we have tried to make as self-contained as possible, excluding only well-known relations, which can be found in the textbooks cited. Here we also include a discussion of the approach to localization using the potential well analogy as developed by Economou and Soukoulis, and its application to amorphous semiconductors.

The homogeneous model treated in Chaps. 4 and 5 is certainly an oversimplification. We therefore include in Chap. 6 a long-range potential. It is shown that this leads to an understanding of the unusual features of the Q function. Another simplification is the tacit assumption that the most relevant energy in transport, the Fermi energy, is entirely temperature independent. From any reasonable density of states model for a-Si : H we conclude that the statistical shift cannot be ignored. In Chap. 7 we show that this effect essentially explains the Meyer–Neldel rule. Furthermore, the shift of the optical band gap is introduced. The implications of the statistical shift on the conductivity, the thermoelectric power, the space-charge limited current and the field effect are discussed in detail.

The theory developed so far is then applied in particular to a-Si : H and a-Ge : H in Chap. 8, while in the concluding Chap. 9 some open questions are presented. It is pointed out that the theory of equilibrium transport in amorphous semiconductors, although considerably developed, is still far from being complete.

1.5 Similar Systems

There is a variety of systems which are, in one or the other respect, similar to amorphous semiconductors. These are, e.g., impurity bands which are used to study localization, hopping conduction and correlation effects. As a testing ground for theories, impurity bands are more suitable than amorphous semiconductors, since they more closely resemble the models treated by theorists. Furthermore, in contrast to amorphous semiconductors, their relevant microscopic parameters are much better known.

The properties of a two-dimensional electron gas in inversion layers or in heterostructures are influenced by disorder. In fact, the quantum Hall effect [v. Klitzing et al., 1980] cannot be understood without taking disorder into account [see, e.g., Ando, 1982; Schweitzer et al., 1985].

In the rapidly growing field of the physics and application of quantum wells and superlattices made from multiple quantum wells, disorder has also to be considered. Localized excitonic states may form due to fluctuations in the well thickness or due to compositional disorder in the barrier material, and energy relaxation of excitons within the inhomogeneously broadened line can be recorded. Theoretical interpretation of the rapidly growing number of experimental data is still in a preliminary state, though the consideration of disorder effects will certainly necessitate more sophisticated theories. On the other hand, superlattices with deliberately introduced disorder [Chomette et al., 1986] can be used to study disorder effects in a well-defined situation.

Molecular crystals are characterized by small electronic band widths and strong excitonic and polaronic effects. The transition from coherent band-like transport to incoherent hopping-like transport can be studied both experimentally and theoretically (for a review see Kenkre and Reineker [1982]). Some theoretical concepts developed in this field can also be applied (with suitable modifications) for amorphous semiconductors.

1.6 Survey of Some Experimental Non-Transport Techniques

In order to investigate the electronic properties of amorphous semiconductors a large number of experimental techniques are used. Standard optical measurements show that the absorption spectra of amorphous semiconductors deviate considerably from those of their respective crystalline modification [Donovan and Spicer, 1968; Beaglehole and Zavetova, 1970] while photoemission spectra yield much less dramatic differences for the density of states spectra [Ley, 1984]. The electroabsorption spectra of amorphous semiconductors are quite unusual and dramatically differ from those of crystalline semiconductors [Weiser et al., 1988]. With the availability of very short laser pulses, transient grating techniques have been used to investigate the relaxation of optically excited carriers down to the sub-picosecond time scale [Noll and Göbel, 1987].

Photoconductivity in amorphous semiconductors is of the utmost technological importance. Various processes are essential both for the photoconductivity and in determining the photoluminescence signal: optical excitation of carriers, energy relaxation within extended states and across the mobility edge, transport close to the mobility edge, trapping in and detrapping from localized band tail states, hopping, and recombination. While the theoretical description of some of these processes is well-developed, the recombination mechanism, in particular, is still poorly understood, although complicated and possibly realistic models do exist [Dersch et al., 1983; Carius and Fuhs, 1985]. A presentation of the present state of this field would require another review and lies outside the scope of this book; however, time-of-flight experiments do probe some of the aforementioned processes and will be treated here. We restrict ourselves, however, to time-scales where recombination can be considered to be unimportant.

The distribution of states within the pseudogap can be studied by a variety of defect spectroscopic methods, including electron-spin resonance, deep-level transient spectroscopy, capacitance measurements, refined optical absorption experiments, field-effect measurements, etc. We shall occasionally make use of information about the density of states distribution gained from these studies.

2. Elementary Treatment of Transport in Amorphous Semiconductors

In this chapter we present the more general properties of amorphous semiconductors. We start with a short discussion of the electronic properties of crystalline tetrahedrally coordinated semiconductors. Some of these properties change surprisingly little if we go from crystalline to amorphous systems. The next sections are devoted to a brief description of the electronic states and of the phonon states in amorphous semiconductors. We then proceed to discuss a simplified model of disorder, the famous Anderson model, and introduce the concept of localization. The scaling theory of localization will provide us with a general framework for the discussion of electronic transport processes in terms of two limiting cases. These are transport in extended states beyond a mobility edge and transport within localized states by hopping. We shall see in Chaps. 4 and 5 which are devoted to the theory that the electron-phonon coupling somehow blurs this distinction. That is why we derive in the last section of this chapter the Kubo–Greenwood formula which will serve as a more general frame for the description of electronic transport processes. In our treatment we shall ignore the electron-electron interaction except for that part that is already included in the mean Hartree–Fock field. In Sect. 7.1 we shall include this interaction in the simplifying picture of Hubbard's [1964] model.

2.1 The Perfect Tetrahedrally Coordinated Lattice

The traditional description of the diamond lattice stressing the long-range order is given in terms of an fcc-lattice with two basis atoms oriented along the (111) direction. Alternatively one may take the chemist's point of view concentrating on the short-range order and can build up the lattice using small clusters of atoms. This approach is more suitable as a starting point for the discussion of an amorphous tetrahedrally coordinated network.

The crystalline diamond lattices of Si and Ge can be considered as composed of tetrahedra attached to each other in the eclipsed position as shown in Fig. 2.1a. All atoms have the same coordination number $Z = 4$. Closed rings of atoms can be found and the minimum number of atoms in a ring is 6. Two bonds attached to a single atom define the bond angle. The distance between nearest neighbours is called the bond length. The dihedral angle, φ, is defined by the relative orientation of two tetrahedra connected by a common bond. The eclipsed (diamond lattice) configuration corresponds to $\varphi = \pi/3$ while the staggered position (of the wurtzite lattice) is given by $\varphi = 0$.

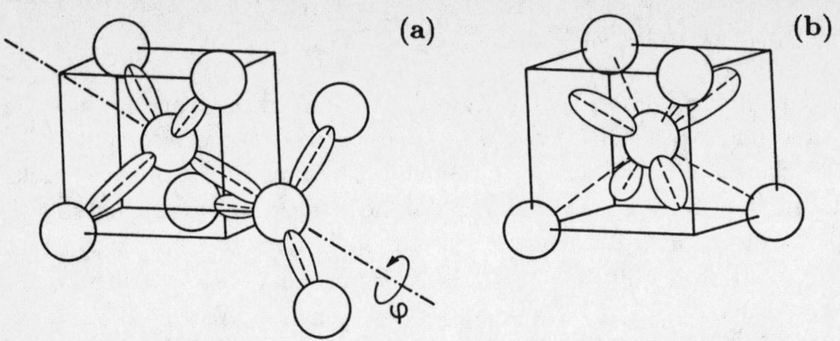

Fig. 2.1. The diamond lattice. (**a**) the bonding orbitals connect nearest neighbours, (**b**) the antibonding orbitals extend into the "holes" of the diamond structure

From standard band structure methods one obtains the band structure, the density of states, the wave functions, and the optical spectra of crystalline semiconductors. The density of states at the band edges rises proportional to $E^{1/2}$ (as is the case in all crystals) and there are no intrinsic bulk states within the fundamental band gap. The valence band edge at the Brillouin zone center has the symmetry Γ'_{25} (if we neglect spin-orbit coupling). The upper part of the valence band is essentially p-like, while the lower part, separated from the upper part by a dip in the density of states, is predominantly s-like. The lowest minimum of the conduction band is along the (111) direction in Ge (symmetry L_1) and along the (100) direction in Si (symmetry X_1). If one circumscribes the atoms by touching spheres most of the valence band charge is found in these spheres [Maschke et al., 1974]. There is a bond charge concentrated along the bonds centered in-between nearest neighbour atoms which is indicated in Fig. 2.1a. In Fig. 2.1b we sketch the charge density for the conduction band minima which is situated predominantly outside the atomic spheres. These states have essentially antibonding character with a minimum of the charge density at the bond center. The lobes extend into the large interstitial "holes" present in the diamond structure.

The optical spectrum is dominated by the prominent peak called E_2 due to direct transitions near the X point in the Brillouin zone. There are two saddle points in the band structure leading to a large peak in the joint density of states. The optical transition matrix element is also large due to Umklapp enhancement which also contributes to the E_2 peak. Both Si and Ge are indirect semiconductors and hence the optical gap for direct transitions is considerably larger than the minimum band separation. The optical band gap, E_g^{opt}, depends on the temperature due to two different mechanisms [Allen and Cardona, 1981]. There is an explicit mechanism caused by the electron-phonon coupling and leading to an approximately linear red shift of the band gap for temperatures above the Debye temperature. The implicit mechanism is connected with the thermal lattice expansion and has a somewhat smaller effect than the explicit mechanism. For elevated temperatures both effects lead to a red shift of the gap that can be approximated by

$$E_g^{opt}(T) = E_g^{opt}(0) - \gamma^{opt}\, T \tag{2.1.1}$$

with $\gamma^{opt} = 470\,\mu eV/K$ for a-Si:H [Fritzsche, 1980].

2.2 The Random Network

The simplest model for an amorphous tetrahydrally bonded semiconductor is the continuous random network (CRN) which was proposed by Zachariasen [1932] for ordinary oxide glasses. This model can be built up by means of balls and sticks starting from tetrahedra with slightly varying bond angles and bond lengths and allowing for arbitrary dihedral angles. Various models have been built by hand [Polk, 1971; Turnbull and Polk, 1972; Polk and Boudreaux, 1973; Steinhardt et al., 1974; Connell and Temkin, 1974; see also Wooten and Weaire, 1987, for a review] or by computer [Shevshik and Paul, 1972; Henderson and Herman, 1972] allowing for structural relaxation in addition. The radial distribution function (RDF) can be calculated for these models and can be compared with experimental data obtained from electron diffraction or x-ray diffraction experiments with real amorphous samples [Moss and Graczyk, 1969, 1970; Temkin et al., 1973; Grigorovici, 1973]. From this comparison one can infer that the bond length variation is rather small with 3%, as is the bond angle variation (about 10%) while the dihedral angle distribution is presumably entirely structureless [Barna et al., 1976; Lannin, 1987] or has at most some broad structure [Barna et al., 1977; Temkin, 1978]. In this particular sense the CRN has a definite (distorted) short range order but no long range order. A very good review on the structural aspects of a-Si : H has been given by Knights [1984].

While the original Polk model started from a seed that contained five-membered and six-membered rings the model of Connel and Temkin [1974] did not allow for odd-membered rings at all. This extra condition did not lead to particular difficulties in the construction of the model nor did it lead to new features in the RDF calculated from the model. It has been shown, however, that odd-membered rings are responsible for filling up the dip in the density of valence states between the upper p states and the lower s states [Joannopoulos and Cohen, 1973].

2.2.1 States Associated with Fluctuations of Bond Lengths, Bond Angles, and Dihedral Angles

Cluster calculations performed for very small clusters containing just one tetrahedron or two connected tetrahedra showed that the lower states in the conduction band are hardly affected by variations in bond length, bond angle, or dihedral angle [Singh, 1981]. The uppermost states of the valence bands, on the other hand, are considerably influenced by a variation of the dihedral angle φ. It seems to be plausible from the last section that any variation of the potential in the interstitial "holes" of a CRN must influence the energies of the antibonding states which form the lowest conduction band states. This suggestion should be confirmed by cluster calculations.

The density of states distribution of the valence bands of amorphous semiconductors has been studied extensively by photoemission using vacuum ultraviolet radiation [Donovan, 1970; Donovan and Spicer, 1968; Ribbing et al., 1971; Pierce and Spicer, 1972] and x-ray sources [Ley et al., 1972; Eastman et al., 1974]. The results [see, e.g., Ley, 1984] differ remarkably little from the corresponding data for the crystalline modification, except that the structures in amorphous semiconductors are significantly broadened. We have already mentioned that in a-Si and a-Ge the lower part of the valence bands is separated by a dip in the density of states from the upper part. This dip is practically absent in the measured photoemission data for the amorphous modification.

Optical absorption measurements for all amorphous semiconductors reveal an exponential absorption edge called the Urbach tail [Urbach, 1953]

$$\varepsilon''(\omega) \sim \exp(\hbar\omega/\Delta E_{\mathrm{opt}}) \tag{2.2.1}$$

for sufficiently small photon energies. The Urbach parameter ΔE_{opt} is close to 50 meV for a-Si : H and a-Ge : H. From time-of-flight experiments (see Sect. 3.6) one deduces an exponential tail for the conduction band. For this tail the slope parameter is about half the value of ΔE_{opt}. The usual assumptions that the optical transition matrix elements are at most weakly energy dependent and that the k-selection rule is totally relaxed, therefore, lead to an exponential tail of the valence band with a slope equal to ΔE_{opt}.

Theoretical work confirms the existence of exponential tails [Halperin and Lax, 1966; Soukoulis et al., 1984; Soukoulis et al., 1985; Economou et al., 1987; Economou and Bacalis, 1987]. Model calculations for the optical properties of disordered semiconductors [Herman and Van Dyke, 1968; Donovan and Spicer, 1968; Tauc, 1969; Brust, 1969; Kramer et al., 1970; Maschke and Thomas, 1970] show that the rather featureless experimental $\varepsilon''(\omega)$ curves [Beaglehole and Zavetova, 1970; Donovan et al., 1970] are well reproduced by a convolution of the valence band density of states, $g_{\mathrm{V}}(E)$, with the conduction band density of states, $g_{\mathrm{C}}(E)$, as proposed by Tauc et al. [1969]

$$n_{\mathrm{conv}}(\omega) = \int_{E_{\mathrm{F}}-\hbar\omega}^{E_{\mathrm{F}}} g_{\mathrm{V}}(E)\, g_{\mathrm{C}}(E+\hbar\omega)\, dE \quad . \tag{2.2.2}$$

There is no prominent influence of the optical transition matrix elements on the shape of the optical absorption edge as long as homogeneous models are considered. For models where a significant long-ranged random Coulomb potential is taken into account the transition matrix elements depend exponentially on the photon energy [Dersch et al., 1987]. In this case the experimental absorption data would not reflect the convoluted density of states.

The width of the band gap depends on the preparation, thermal treatment, and in particular on the hydrogen content of the amorphous films. Since alloying of, e.g., a-Si : H with germanium or carbon presents no difficulties at all one has a material for which the band gap can be controlled in a wide range.

2.2.2 Intrinsic Defects

Phillips [1979] has pointed out that it would be impossible to construct an infinite CRN without extremely large internal strain and stress. Broken bonds or unsaturated bonds will, therefore, be formed to release the internal tension.

These intrinsic defects will produce defect states between the bonding and the antibonding states.

The most prominent intrinsic defect is the unsaturated bond, called the dangling bond or T_3 center (the name indicates a threefold coordinated atom). Dangling bonds manifest their presence by an ESR signal with a g value $g = 2.0055$ [Brodsky and Title, 1969; Stuke, 1976]. For unhydrogenated a-Si this spin density can be as large as 10^{19} cm^{-3}, whereas for hydrogenated material the spin density can be below the limit of detectability (which is about 10^{15} cm^{-3}). The figures indicate the dramatic reduction in the density of pseudogap states that can be obtained by hydrogenation. Note, however, that a reduction of the spin density does not mean that there are less defects: the energetic position of the hydrogenated defects is deep in the valence bands and these states, therefore, do not manifest themselves by states in the pseudogap.

Pantelides [1986] has recently suggested that five-fold coordinated T_5 atoms (called floating bonds) should also exist. He considers these floating bonds as the defect which produces the $g = 2.0055$ spin resonance signal while the dangling bond states are supposed to have a negative effective correlation energy (see Sect. 7.1), in agreement with theoretical calculations [Bar-Yam et al., 1986]. Interconversion between T_3 and T_5 centers should be highly probable which can explain the variation in spin density observed by annealing and by light soaking [Hauschildt et al., 1982]. Note, however, that we know from light induced ESR and from dark ESR that there is only one single defect in the pseudogap of a-Si : H with a concentration exceeding 5×10^{14} cm^{-3} [Fritzsche, 1987].

Amorphous solids are in a metastable state. It has been shown recently [Kakalios and Street, 1986; Street et al., 1987] that intrinsic defects are created at elevated temperatures ($T \geq 410$ K) in thermal equilibrium with the occupancy of the tail states by electrons and holes. Part of this defect structure can be frozen in if the sample is quenched to temperatures below 410 K.

2.2.3 Inhomogeneous Structure

For amorphous films prepared by evaporation or by sputtering one quite generally observes an inhomogeneous structure, called the supernetwork or columnar structure with a typical lateral size of about 100 Å [Donovan and Heinemann, 1971]. This structure can be described as columns of high density (close to the crystalline density) surrounded by an interconnected network of low-density ("tissue") material [Barna et al., 1974]. If the deposition is performed under high vacuum conditions this structure can be changed by annealing: The tissue material is transformed into high density material and the density deficit is balanced by the formation of voids. If, on the other hand, the sample is contaminated by foreign atoms the supernetwork structure is stabilized.

It has been demonstrated [Knights and Lujan, 1979] that under special deposition conditions samples prepared by the glow-discharge decomposition of silane also show a supernetwork structure. It is, however, generally accepted now that samples prepared under "optimized" conditions are structurally homogeneous on a mesoscopic scale. This has been shown by small angle scatttering for some samples at least [Barna et al., 1976; Leadbetter et al., 1980, 1981]. It is, however, not possible to rule out the existence of microvoids. Several authors [Schiff et al., 1981; Beyer et al., 1981; Yamasaki et al., 1982] report that heavy doping by boron can lead to drastic microstructural inhomogeneities.

2.2.4 Extrinsic Defects and Impurities

In hydrogenated amorphous semiconductors hydrogen itself is of course the most numerous impurity. In a similar way as in crystalline silicon this impurity does not introduce localized states in the gap. Instead hydrogen serves primarily to passivate the defects already present in the amorphous system and thus to decrease the density of defect states in the pseudogap [Brodsky et al., 1970]. It has been shown that in a-Si:H the hydrogen atoms are present in two different configurations: NMR experiments [Reimer et al., 1980, 1981] show that about 4 atomic percent of hydrogen can be incorporated as dilute hydrogen while the rest is found in clusters of 5 to 7 hydrogen atoms [Baum et al., 1986]. A similar inhomogeneous distribution of the hydrogen atoms is also observed for microcrystalline Si:H [Kumeda et al., 1983].

In order to control the electronic properties one is interested to incorporate atoms that can act as donors and acceptors, respectively. Since the pioneering work of Spear and LeComber [1975] the most popular doping method for a-Si:H consists in the admixture of PH_3 or B_2H_6 gas to the silane prior to decomposition. Doping is also possible by ion implantation [Müller et al., 1977; Spear et al., 1979; Kalbitzer et al., 1980] and leads to both substitutional and interstitial donors and to substitutional acceptors. Beyer and Fischer [1977] have shown that interstitial doping by in-diffusion of Lithium is also possible.

The electrical activity of the dopant atoms is related to a complex reaction with both dangling bonds and hydrogen. This mechanism is controlled at least in part by the position of the Fermi energy at the growing surface [Street et al., 1981; Stutzmann and Street, 1985] and interferes with the metastable dangling bonds. The experiments show that the density of midgap states is severely increased upon doping. There is at present no satisfactory model that describes the physics of these doping reactions although it would be of great interest for the understanding of the physics and also for optimization purposes.

Besides the intentional incorporation of dopants several other extrinsic defects can be present as contaminants. SIMS experiments show [Fritzsche, 1987] that about 10^{19} cm^{-3} defects like oxygen, nitrogen, or carbon are present in a-Si:H films prepared under the usual conditions.

2.3 Phonons

Phonons can be observed spectroscopically by infrared absorption and by Raman spectroscopy. In a crystalline solid the lattice periodicity gives rise to very strict selection rules for these transitions. It is, therefore, possible to observe the lines of a few phonon modes at wavevectors close to zero only. In amorphous semiconductors there is no long-range order and, therefore, the selection rules will be relaxed [Lucovsky, 1974]. In fact, from Raman and infrared absorption data [Smith Jr. et al., 1971; Wihl et al., 1972; Alben et al., 1975] one obtains spectra that very much resemble the broadened crystalline one-phonon density of states (if in case of the Raman spectra the proper phonon occupation numbers are extracted from the spectra). There is surprisingly little influence from a possible energy dependent transition matrix element.

Local modes are also present and in particular prominent in hydrogenated material [Brodsky et al., 1977; Knights et al., 1978; Cardona, 1983; for a recent review see Zanzucchi, 1984] which can be attributed to the various bond bending (rocking and wagging) and bond stretching modes of the Si : H bond. Various bonding configurations such as SiH, SiH_2, and SiH_3 bonds can be identified [Lucovsky and Hayes, 1979].

In this book we are interested in phonons mainly in connection with the electron-phonon coupling. Here we always need a weighted average over all phonon modes. We shall keep in mind that of course the phonons can no longer be labelled by a wave vector and that we should use some other quantum number instead. For simplicity we shall keep the symbol $\boldsymbol{q}$ as the phonon quantum number which, at least for acoustical phonons in the limit of long wavelengths, makes some sense. We shall in particular exploit the fact that the density of states for the extended phonons closely resembles the corresponding crystalline density of states.

2.4 The Anderson Model

In the previous section we have given a rather complicated picture of disorder in amorphous semiconductors due to structural disorder, intrinsic, and extrinsic defects and to inhomogeneities. In order to understand the effect of disorder we shall now concentrate on a much simpler model for a disordered system as presented by Anderson [1958] in his classic paper.

This model starts with a crystal in a single band approximation using the Wannier (or site) representation. The Hamiltonian can be written as

$$H = \sum_{i} \varepsilon_i \, n_i + \sum_{i \neq j} J_{ij} \, c_i^{\dagger} \, c_j \tag{2.4.1}$$

where the first sum includes all the atomic eigenvalues at the positions ("sites") $\boldsymbol{R}_i$ in the crystalline lattice while the second term sums up all transfer matrix elements. The $c_i^{\dagger}$ and c_j are the creation (annihilation) operators for fermions in Wannier states $|i\rangle$ and $|j\rangle$, respectively, centered around $\boldsymbol{R}_i$ and $n_i = c_i^{\dagger} \, c_i$. While the Hamiltonian (2.4.1) describes all possible crystals (in a one-band approximation however) the essential physics of the Anderson model is retained if we cast the model in a simpler form where

$$J_{ij} = \begin{cases} J & i,j \text{ are nearest neighbours} \\ 0 & \text{else} \end{cases} \quad . \tag{2.4.2}$$

In a crystal the energies ε_i are independent of site index and just give the center of the band determined by (2.4.1). For a disordered system the values of ε_i are taken to be random numbers. In the simplest and most thoroughly studied version of the Anderson model the ε_i are taken from a uniform random number distribution of width W. For $W = 0$ we retain a crystal, the electronic states form a band of band width $B = 2 \cdot Z \cdot J$ where Z is again the coordination number of the lattice and the eigenstates are Bloch states which can be labelled by a wave vector $\boldsymbol{k}$ according to

$$\Psi_{\boldsymbol{k}} = \frac{1}{\sqrt{N}} \sum_i \mathrm{e}^{\mathrm{i}\boldsymbol{k}\cdot\boldsymbol{R}_i} \, |i\rangle \tag{2.4.3}$$

(N denotes the total number of sites). For $W \neq 0$ but small compared to J the density of states at the band edges rises no longer as the square root of the energy. Instead, tails are formed at both band edges which are composed of localized states, i.e., of states for which the wavefunction decays exponentially for large distances from some center. Any disorder, however small, produces some localized states at the band edges. These form a tail of finite width if the distribution of the ε_i also has a finite width. Figures that display the charge density of localized and of delocalized states for a two-dimensional Anderson model are shown by Yoshino and Okazaki [1977].

The states in the center of the band are (for moderate disorder W/J) far less affected by the disorder than the states in the tails. These states are, therefore, delocalized but of course no longer Bloch states. It is often assumed [Mott, 1967; Hindley, 1970; Friedman, 1971] that these states can be represented as

$$\Psi_\lambda = \frac{1}{\sqrt{N}} \sum_i a_i^\lambda \, |i\rangle \tag{2.4.4}$$

where the a_i^λ are of order unity but with random phases (note, however, that for one-dimensional systems Moullet et al. [1985] have shown that these phases are far from being random). For small disorder there are two distinct energies, the mobility edges, E_{C} and E'_{C}, respectively, [Bányai, 1964; Mott, 1966], that separate the localized from the delocalized states as indicated in Fig. 2.2. With increasing disorder both mobility edges move towards the center of the band. The position of the mobility edges as a function of W/J has been investigated by

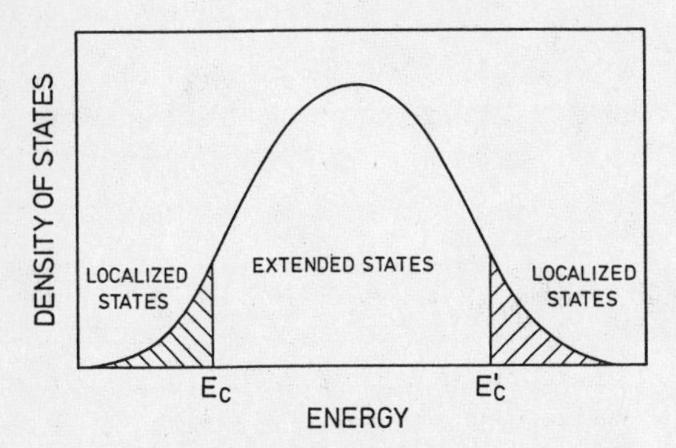

Fig. 2.2. Density of states distribution for the Anderson Hamiltonian

several groups [Economou and Cohen, 1970, 1972; Schönhammer, 1971; Abou-Chacra and Thouless, 1974]. If the disorder energies exceed the width of the crystalline band considerably (for $W/J \cong 16$ in simple cubic lattices according to MacKinnon and Kramer [1981]) all states are localized. We shall in the following treat the case that the majority of the states are delocalized and only states in the tails are localized.

2.4.1 The Minimum Metallic Conductivity

In order to investigate the transport properties of a system with a mobility edge separating delocalized states from localized states we shall in the next two subsections consider the case of a degenerate gas of noninteracting electrons at zero temperature. Let us assume that we have fixed the disorder (or rather W/J) and that we can move the Fermi energy above or below the mobility edge E_C. For E_F below E_C the conductivity is zero because all states are localized whereas for $E_F > E_C$ there will be a finite conductivity. The question to be asked is if this conductivity can be arbitrarily small.

It has been pointed out by Ioffe and Regel [1960] that electrons can be described as free particles with a mean free path l_{free} only if this mean free path is sufficiently large. If we use the dc Drude formula for a degenerate electron gas

$$\sigma = \frac{e^2\, k_F^2\, l_{\mathrm{free}}}{3\, \pi^2\, \hbar} \tag{2.4.5}$$

where k_F is the Fermi wave number we must make sure that at least

$$k_F\, l_{\mathrm{free}} \geq 1 \tag{2.4.6}$$

or the concept of free particles loses its meaning. Mott [1967] has argued that the Drude formula can be safely used as long as (2.4.6) holds. He gives reasons why the conductivity cannot be smaller than that given by $k_F\, l_{\mathrm{free}} = 1$. Thus, if the mean free path becomes smaller the conductivity must drop to zero. This means that for a metallic system with a minimum mean free path l_{min} the conductivity at the mobility edge will be given by

$$\sigma_M = \frac{e^2}{\hbar}\, \frac{1}{l_{\mathrm{min}}}\, C_3 \tag{2.4.7}$$

with $C_3 \cong 0.03$. This is the famous minimum metallic conductivity which has kept the field of localization theories alive for some 20 years [see Mott and Davis, 1971, 1979; Mott, 1974, 1987].

2.4.2 The Scaling Theory of Localization

The general behaviour of the conductivity close to the mobility edge can be studied by scaling theory. This theory is based on ideas put forward by Thouless [Thouless, 1974, 1980; Liciardello and Thouless, 1975,1978] that the relevant quantity to study should be the conductance, $G(L)$, of a cube of linear length

L rather than its conductivity, σ. Since for conductances there is a natural unit $e^2/\hbar$ we shall consider the dimensionless quantity $g(L) = G(L)\,\hbar\,/e^2$. The basic idea of Thouless was that upon increasing the size of a system, the system conductance will scale in a peculiar way. This idea was developed by the "gang of four" [Abrahams, Anderson, Liciardello, and Ramakrishnan, 1979; see also the review by Lee and Ramakrishnan, 1985] to a scaling theory. We shall treat here the case of three-dimensional systems exclusively, although the cases of other dimensionality have their own interest.

Consider again a degenerate gas of noninteracting electrons in a cube of length l at zero temperature. This length shall be chosen microscopically small but larger than the minimum length scale where diffusion (i.e., scattering limited motion) can be defined. We refer to

$$g_0 = g(l) \tag{2.4.8}$$

as the microscopic conductance. For highly conducting systems (g_0 large) we will have Ohm's law which means that the conductance scales with the cube size as

$$g(L) = \frac{\hbar}{e^2}\,\sigma\,L \tag{2.4.9}$$

as long as L is much larger than l. For poorly conducting systems (g_0 small) the states around the Fermi energy will be localized with a localization length ξ_{loc}. The conductance will, therefore, decrease with increasing size as

$$g(L) = \tilde{g}\exp(-L/\xi_{\mathrm{loc}}) \tag{2.4.10}$$

with some constant prefactor $\tilde{g}$, which is clearly nonohmic. We thus see that depending on the microscopic conductance g_0 on a length scale l the scaling of the conductance with sample size is entirely different. The "gang of four" argues that the logarithmic derivative

$$\beta(g) = \frac{d\,\ln(g)}{d\,\ln(L)} = \frac{L}{g}\,\frac{dg}{dL} \tag{2.4.11}$$

is a function of the conductance alone (and does, in particular, not depend on L). Furthermore the basic assumption is that this function is single-valued, monotonous, and continuous. Since $\beta(g)$ is known for the extreme cases of very large and very small values of g

$$\beta(g) = 1 \qquad g \to \infty \tag{2.4.12}$$

and

$$\beta(g) = \ln(g/\tilde{g}) \qquad g \to 0 \tag{2.4.13}$$

we obtain a scaling function as schematically given in Fig. 2.3.

Suppose that we have a small system (the length must be at least l of course) for which the conductance g_0 is larger than the value g_3, which is the zero of the β function. If we increase the size of this system the conductance will increase owing to the scaling relation. If we iterate the size doubling we will always increase the system conductance and, finally, reach the Ohmic region. If, on the other hand, we started with a conductance below g_3, the scaling function tells us that the conductance decreases and that this continues until for very large systems the conductance tends to zero. Since g_0 depends on E_F,

$$g_0(E_F = E_C) = g_3 \tag{2.4.14}$$

is the microscopic conductance for the mobility edge E_C. The corresponding conductivity, $e^2 g_3/\hbar l$, is the smallest value of the microscopic conductivity which leads upon scaling up to the ohmic behaviour, described by (2.4.12). It can, therefore, be identified with Mott's minimum metallic conductivity, σ_M.

Information about the behaviour of the conductivity at macroscopic length scales L can be extracted from the scaling function. We consider a system with a conductance g_0 slightly above g_3 at the length l, i.e., the Fermi energy is slightly above E_C. We approximate the scaling function by (see Fig. 2.4)

$$\beta(g) = \frac{1}{\lambda}\ln(g/g_3) \qquad \text{for} \qquad g_3 \le g \le g_1 = g(l_1) \tag{2.4.15}$$

and

$$\beta(g) = 1 \qquad \text{for} \qquad g \ge g_1 \tag{2.4.16}$$

with some slope $1/\lambda$. We expand in particular

$$\beta(g_0) = \frac{1}{\lambda}\left[\frac{g_0 - g_3}{g_3}\right] = \frac{\delta g}{\lambda} \quad . \tag{2.4.17}$$

From geometry we have

$$\ln(g_1/g_3) = \lambda \tag{2.4.18}$$

and from the linear approximations we find that

$$\frac{d\,\ln(g/g_3)}{\ln(g/g_3)} = \frac{1}{\lambda}\,d\,\ln(L) \quad . \tag{2.4.19}$$

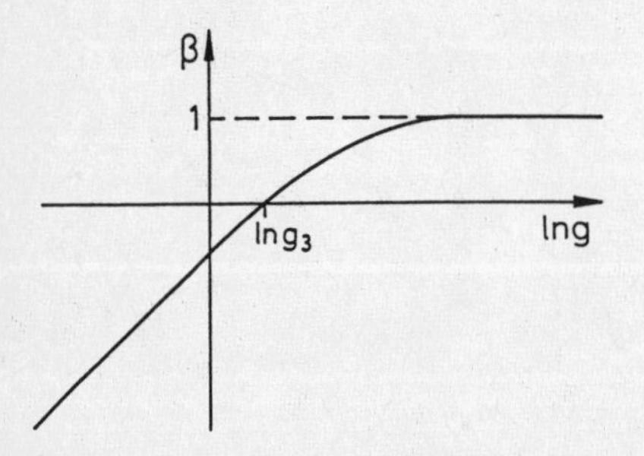

Fig. 2.3. The scaling function $\beta(g)$ for a threedimensional system

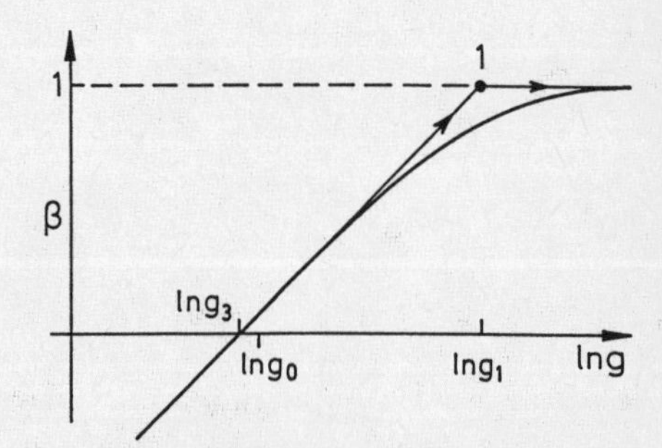

Fig. 2.4. The integration path used for the approximate determination of the conductivity

We integrate this latter equation starting with g_0

$$\int_{\ln(g_0/g_3)}^{\ln(g_1/g_3)} \frac{du}{u} = \frac{1}{\lambda} \ln(l_1/l) \tag{2.4.20}$$

which is equivalent to

$$l_1 = l \left[\frac{\lambda}{\delta g}\right]^\lambda \tag{2.4.21}$$

because of (2.4.17) and (2.4.18). Integrating from l_1 to L (with $\beta(g) = 1$) we find

$$g = g_1 \frac{L}{l_1} \tag{2.4.22}$$

which in combination with (2.4.18) leads us to the final result

$$\sigma = \frac{e^2}{\hbar} A \frac{g_3}{l} (\delta g)^\lambda \qquad \text{with} \qquad A = \left(\frac{\mathrm{e}}{\lambda}\right)^\lambda \quad . \tag{2.4.23}$$

We thus see that the conductivity of a macroscopic system depends on the conductance of the corresponding system at the length scale l. We do not obtain a minimum metallic conductivity. Instead, if we expand the microscopic conductance g_0 of the system at length scale l as a function of the position of the Fermi energy

$$g_0(E_\mathrm{F}) = g_3 + (E_\mathrm{F} - E_\mathrm{C})\, g_0'(E_\mathrm{C}) \tag{2.4.24}$$

we find that the conductivity is proportional to $(E_\mathrm{F} - E_\mathrm{C})^\lambda$. Several microscopic theories ([Belitz and Götze, 1983; Vollhardt and Wölfle, 1982]; see also Chap. 4) give $\lambda = 1$ which corresponds to $A = \mathrm{e}$. Measuring E_C from the lower band edge we obtain with

$$g_0'(E_\mathrm{C}) = g_3/E_\mathrm{C} \tag{2.4.25}$$

$$\sigma = \frac{e^2}{\hbar} \frac{\mathrm{e}\, g_3}{l} \frac{E_\mathrm{F} - E_\mathrm{C}}{E_\mathrm{C}} \quad . \tag{2.4.26}$$

It is customary to write

$$\sigma = \frac{e^2}{\hbar} \frac{g_3}{\xi} \tag{2.4.27}$$

defining the correlation length ξ which diverges as E_F approaches the mobility edge E_C. For $\lambda = 1$ we find

$$\xi = \frac{E_\mathrm{C}}{E_\mathrm{F} - E_\mathrm{C}} \frac{l}{\mathrm{e}} \quad . \tag{2.4.28}$$

It can be shown easily using (2.4.10) and $\lambda = 1$ that $\tilde{g} = g_3$ and that $\xi_{\rm loc}$ also diverges with the same exponent λ if we approach the mobility edge from the localized side.

We see that for disordered systems at zero temperature there is no minimum metallic conductivity according to the scaling theory. Instead, the conductivity is continuous at the mobility edge. For semiconductors at zero temperatures the conductivity is zero anyway. But if, e.g., electrons are injected into the conduction band of an amorphous semiconductor we always have inelastic processes (even at zero temperature) which lead to an inelastic scattering length $L_{\rm i}$. If in our present model inelastic processes are included the conductivity at $E_{\rm C}$ will, therefore, be given [Thouless, 1977; Imry, 1980] by (2.4.27) with ξ replaced by $L_{\rm i}$

$$\sigma = \frac{e^2}{\hbar} \frac{g_3}{L_{\rm i}} \quad . \tag{2.4.29}$$

(Mott [1988] argues that in the vicinity of g_3 the exponent λ should be zero. For this case he also obtains (2.4.29) and $g_3 \simeq 0.03$.) Unfortunately, the scaling theory does not provide us with a value for σ and also does not give $\sigma(E_{\rm F})$ for $E_{\rm F}$ outside the critical region (which may be small). It will be necessary, therefore, to develop a theory that gives $\sigma(E)$ for all energies and also includes interactions with phonons in order to describe transport in disordered semiconductors with a microscopic theory. This will be done in Chaps. 4 and 5.

2.5 Elementary Transport Model for Amorphous Semiconductors

Before we can apply the results of the previous section to amorphous semiconductors we must amend the single-band model to at least a two-band model. We shall assume that the disorder parameters for both bands will not be drastically different and, therefore, obtain a density of states distribution (the Cohen, Fritzsche, Ovshinski (CFO) model [1969]) as given by Fig. 2.5: localized states extend from the valence band and from the conduction band into the "pseudogap" and overlap in the center of the pseudogap. The neutrality of the system is guaranteed if the Fermi energy separating occupied from unoccupied states is in the overlap region of the tails. Valence band tail states above $E_{\rm F}$ and conduction band tail states below $E_{\rm F}$ are charged in this model. States between the mobility edges at $E_{\rm V}$ and $E_{\rm C}$, respectively, are localized while states beyond these mobility edges are extended. For most glasses the density of states in the center of the pseudogap is either extremely small or zero, thus rendering the material transparent, highly transparent in fact if the defect density can be made sufficiently small.

A more realistic density of states distribution is shown schematically in Fig. 2.6. Here we have also included the intrinsic and extrinsic defects discussed in Sects. 2.2.2 and 2.2.4, respectively. In this figure the Fermi level drawn corresponds to the case of a perfectly compensated sample. In general the density

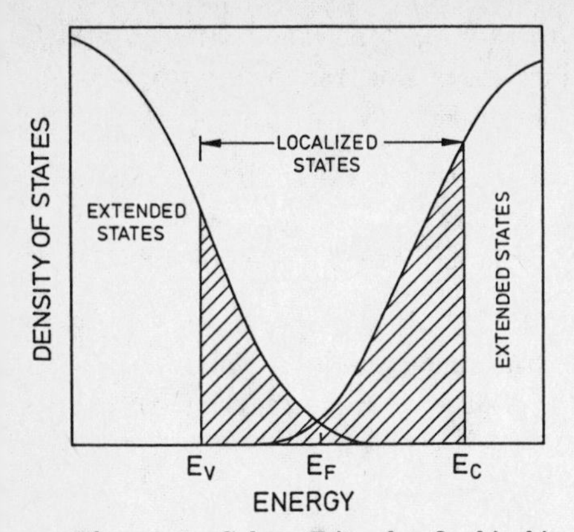

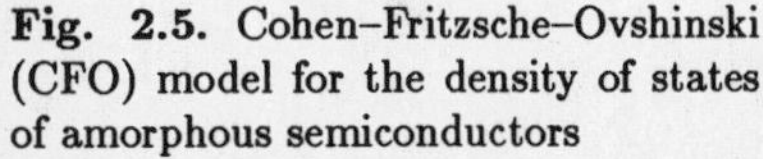

Fig. 2.5. Cohen–Fritzsche–Ovshinski (CFO) model for the density of states of amorphous semiconductors

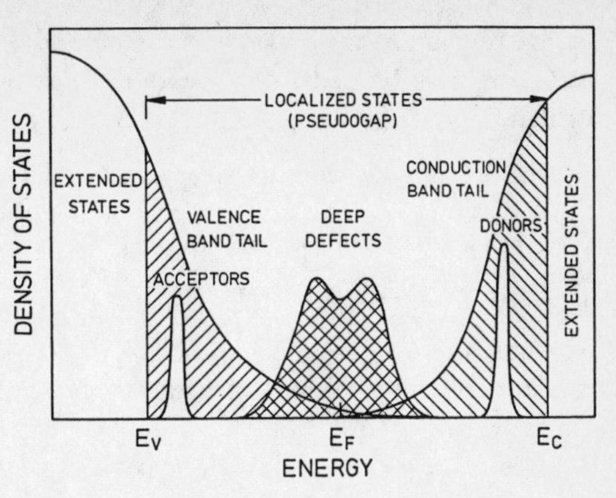

Fig. 2.6. Schematic density of states distribution in the pseudogap for a real amorphous system

of donors will not exactly balance the density of acceptors. We, therefore, expect that the Fermi level will be situated in the upper half (lower half) of the pseudogap for n-type (p-type) samples.

We shall restrict the discussion to the most simple case of an isotropic homogeneous semiconductor. From irreversible thermodynamics one can quite generally derive a relation between the electronic current density $\boldsymbol{j}$ and the concomitant heat current $\boldsymbol{w}$ as a function of the driving forces. This relation is valid as long as the current densities are linear functions of the forces. As driving forces we have the external electric field $\boldsymbol{F}$, a possible gradient of the Fermi energy, E_{F}, and a temperature gradient. The general linear relation reads for carriers of charge q [Onsager, 1931]

$$\boldsymbol{j} = L_{11}\left[\boldsymbol{F} - \frac{1}{q}\,T\,\boldsymbol{\nabla}\left(\frac{E_{\mathrm{F}}}{T}\right)\right] + L_{12}\,T\,\boldsymbol{\nabla}\left(\frac{1}{T}\right) \tag{2.5.1a}$$

$$\boldsymbol{w} = L_{21}\left[\boldsymbol{F} - \frac{1}{q}\,T\,\boldsymbol{\nabla}\left(\frac{E_{\mathrm{F}}}{T}\right)\right] + L_{22}\,T\,\boldsymbol{\nabla}\left(\frac{1}{T}\right) \quad . \tag{2.5.1b}$$

For anisotropic systems the coefficients L_{ij} are tensors. For amorphous semiconductors we shall assume all properties to be isotropic, so the L_{ij} are scalars that depend on the properties of the material and also explicitly on the temperature. In the presence of a magnetic field $\boldsymbol{B}$, however, the coeeficients L_{ij} are no longer scalars although the system is isotropic. There is a symmetry relation (Onsager's reciprocity relation, [Onsager, 1931])

$$L_{12}(\boldsymbol{B}) = L_{21}(-\boldsymbol{B}) \tag{2.5.2a}$$

which in the absence of a magnetic fields becomes the simple scalar relation

$$L_{12} = L_{21} \tag{2.5.2b}$$

(Onsager's symmetry) for a homogeneous system.

In the following sections we shall evaluate the L_{ik} for two limiting cases, transport in extended states beyond the mobility edges and transport within the localized states in the pseudogap. At the present stage we shall assume that

i) the system is homogeneous, i.e., the disorder has a typical length scale which is of the order of a nearest atomic neighbour distance
ii) all energy levels, the Fermi level, and also the mobility edges are not affected by a variation in temperature.

We shall see in Chaps. 6 and 7 that these assumptions cannot be valid in general but must be relaxed in order to obtain a qualitative understanding of the experimental data.

2.6 The Standard Transport Model

We assume that electrons and holes in localized states within the pseudogap do not contribute at all to the electronic transport. Instead, electronic transport is entirely due to carriers in extended states beyond the mobility edges. For simplicity we shall assume that carriers of one polarity only contribute to the transport.

The density of mobile electrons in extended states above the mobility edge E_C is given by an integral over the density of states distribution (a factor of two enters because of the two spin orientations)

$$n_C = 2 \int_{E_C}^{\infty} g(E)\, f_F(E, E_F, T)\, dE \tag{2.6.1}$$

with the Fermi–Dirac distribution function ($\beta = 1/kT$)

$$f_F(E, E_F, T) = \{1 + \exp[\beta(E - E_F)]\}^{-1} . \tag{2.6.2}$$

For holes below the mobility edge E_V we have

$$n_V = 2 \int_{-\infty}^{E_V} g(E)\, [1 - f_F(E, E_F, T)]\, dE . \tag{2.6.3}$$

Since the Fermi energy is usually separated by several kT from the mobility edges we can replace the Fermi distribution functions by the Boltzmann distribution which for electrons reads

$$f_0^e = \exp[-\beta(E - E_F)] \tag{2.6.4}$$

while for holes we replace $1 - f_F(E, E_F, T)$ by

$$f_0^h = \exp[-\beta(E_F - E)] . \tag{2.6.5}$$

These distribution functions describe the case of nondegenerate statistics. Usually one can safely assume that the density of states distribution $g(E)$ above the mobility edge does not change significantly on a kT scale. The use of nondegenerate statistics leads in this case to

$$n_C = 2\,g(E_C)\,kT\,\exp[-\beta(E_C - E_F)] \tag{2.6.6}$$

for the electron density in extended states (for holes the formula is mutatis mutandis the same). The mobile electrons in the extended states perform a random thermal motion. If a small external electric field $\boldsymbol{F}$ is applied there will be a small directed motion parallel (in fact antiparallel for electrons owing to their negative charge $q = -e$) to the field superimposed on the random motion. For this directed component of the motion one defines the mobility, μ_0

$$|\boldsymbol{v}| = \mu_0\,|\boldsymbol{F}|\ . \tag{2.6.7}$$

For simplicity we shall assume that the mobility is roughly the same for all the carriers in the extended states (and zero for all electrons in localized states). Due to the competition between the random thermal motion and the force due to the external field one usually finds that

$$\mu_0 \sim 1/T\ . \tag{2.6.8}$$

We shall refer to this transport model in the following as the *Standard Transport Model.*

2.6.1 dc Conductivity

Because amorphous tetrahedrally bonded semiconductors can be prepared as thin films only most dc conductivity experiments are performed in a gap cell configuration sketched in Fig. 2.7. The film is deposited on a suitable substrate and both current and applied voltage are measured over the gap formed by two parallel electrodes on top of the film. From (2.5.1a)

$$\boldsymbol{j} = \sigma\,\boldsymbol{F} \tag{2.6.9}$$

where σ is the dc conductivity identified with L_{11}. According to our Standard Transport Model we find (for electrons)

$$\boldsymbol{j} = n_C\,q\,\boldsymbol{v} = n_C\,|q|\,\mu_0\,\boldsymbol{F} \tag{2.6.10}$$

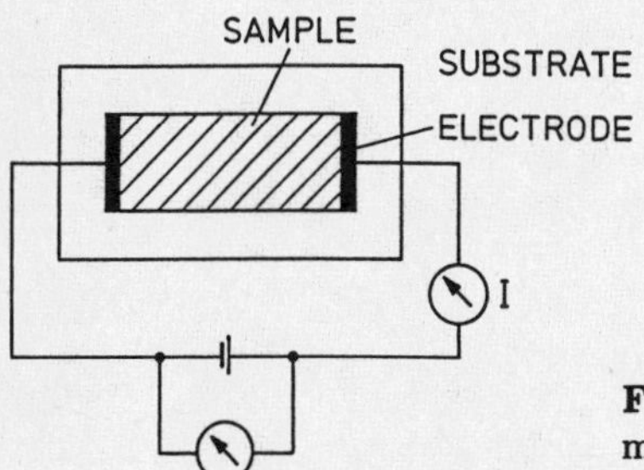

Fig. 2.7. Experimental setup for the conductivity measurement in a gap configuration

and, therefore,

$$\sigma = |q|\,\mu_0\,N_C\,\exp[-\beta(E_C - E_F)] \tag{2.6.11}$$

with the effective density of extended states

$$N_C = 2\,g(E_C)\,kT\ . \tag{2.6.12}$$

This relation is frequently rewritten as

$$\sigma(T) = \sigma_0\,\exp[-\beta(E_C - E_F)] \tag{2.6.13}$$

with the prefactor

$$\sigma_0 = \mu_0\,|q|\,N_C\ . \tag{2.6.14}$$

In our Standard Model the "prefactor" σ_0, which will be referred to as the microscopic prefactor of the conductivity, is temperature independent. If we plot the conductivity on a logarithmic scale versus inverse temperature — this plot is called the "Arrhenius plot" of the conductivity — we obtain a straight line of slope E_σ/k with $E_\sigma = E_C - E_F$ as "activation energy" and an intercept at $1/T = 0$ which is equal to σ_0 (actually the intercept is $\log(\sigma_0)$ on a logarithmic scale). If we move the Fermi energy (by doping, say) we expect that the slope E_σ changes whereas the intercept will be entirely unaffected as is plotted schematically in Fig. 2.8.

For holes we find correspondingly

$$\sigma(T) = \sigma_0\,\exp[-\beta(E_F - E_V)] \tag{2.6.15}$$

with

$$\sigma_0 = q\,\mu_0\,N_V \tag{2.6.16}$$

and

$$N_V = 2\,g(E_V)\,kT\ . \tag{2.6.17}$$

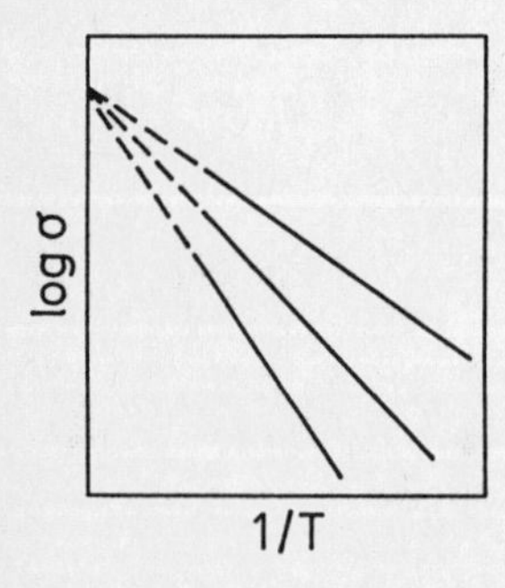

Fig. 2.8. Arrhenius plot for the conductivity in extended states as predicted from the Standard Transport Model

We do not expect the microscopic prefactors, σ_0 , for electrons and holes to be different. In this case a comparison of (2.6.13) to (2.6.15) shows that the position of the Fermi energy within the pseudogap determines predominantly the sign of the majority carriers.

2.6.2 Peltier Coefficient and Thermoelectric Power

Electronic currents are accompanied by heat currents even under isothermal conditions. In order that Onsager's symmetry holds it is necessary that the heat current is normalized properly, i.e., the origin of the energy scale coincides with the Fermi energy. This choice would be rather unpractical in cases where the Fermi energy itself is altered (with temperature, e.g.). We, therefore, leave the origin of the energy scale unspecified and define a properly normalized heat current density, $\tilde{\boldsymbol{w}}$, which under isothermal conditions reads (2.5.1b)

$$\tilde{\boldsymbol{w}} = L_{21}\,\boldsymbol{F} - \frac{\boldsymbol{j}}{q}\,E_{\mathrm{F}} \tag{2.6.18a}$$

$$= \left[\frac{L_{21}}{L_{11}} - \frac{E_{\mathrm{F}}}{q}\right] \cdot \boldsymbol{j} \tag{2.6.18b}$$

$$= \Pi \cdot \boldsymbol{j} \tag{2.6.18c}$$

where the last relation defines the Peltier coefficient, Π.

A quantitative measurement of the Peltier coefficient is difficult because heat currents cannot be measured directly. Instead one measures the difference between the Peltier heat of the sample and that of the lead material which is picked up at one electrode and released at the other electrode. According to our Standard Transport Model we have

$$q\,\tilde{\boldsymbol{w}} = \boldsymbol{j}\,(|E_{\mathrm{m}} - E_{\mathrm{F}}| + kT) \tag{2.6.19}$$

where E_{m} is the energy of the mobility edge for electrons or holes, respectively. In cases where the mobility (or the density of states) is not independent of the energy of the carriers on a kT scale one finds instead

$$q\,\tilde{\boldsymbol{w}} = \boldsymbol{j}\,(|E_{\mathrm{m}} - E_{\mathrm{F}}| + A\,kT) \tag{2.6.20}$$

and, therefore

$$\Pi = \frac{1}{q}\,(|E_{\mathrm{m}} - E_{\mathrm{F}}| + A\,kT) \tag{2.6.21}$$

where A, the "heat of transport" term, is of order unity.

Instead of a measurement of the Peltier coefficient it is more convenient to perform a measurement of the thermoelectric power ("thermopower", "Seebeck coefficient") as sketched in Fig. 2.9: a temperature gradient is established between the electrode gap and the open circuit voltage, ΔV, is measured which defines the Seebeck coefficient, S, via

$$\Delta V = S\,\Delta T\ . \tag{2.6.22}$$

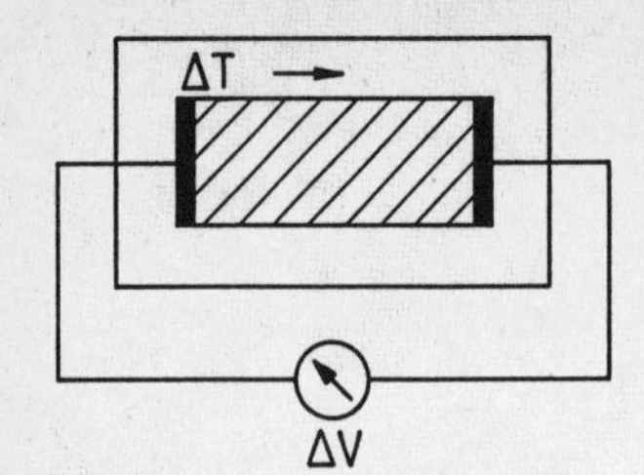

Fig. 2.9. Schematic experimental setup for the determination of the thermoelectric power

From (2.5.1a) we obtain with $\boldsymbol{j} = 0$

$$L_{11}\left[\boldsymbol{F} - \frac{T}{q}\,\nabla\!\left(\frac{E_{\mathrm{F}}}{T}\right)\right] = -L_{12}\,T\,\nabla\!\left(\frac{1}{T}\right) \tag{2.6.23}$$

or

$$\boldsymbol{F} = \left(\frac{L_{12}}{L_{11}} - \frac{E_{\mathrm{F}}}{q}\right)\frac{\nabla T}{T}\ . \tag{2.6.24}$$

By comparison with (2.6.22) we find

$$S = \frac{1}{T}\left(\frac{L_{12}}{L_{11}} - \frac{E_{\mathrm{F}}}{q}\right) \tag{2.6.25}$$

and with the help of Onsager's symmetry relation

$$S = \frac{\Pi}{T} \tag{2.6.26}$$

which in our Standard Transport Model reduces to

$$q\,S = \left(\frac{E_{\mathrm{m}} - E_{\mathrm{F}}}{T} + A\,k\right) \tag{2.6.27}$$

with $A \cong 1$. We therefore, expect from our Standard Model that a plot of the Seebeck coefficient versus inverse temperature, an "Arrhenius plot" of the thermoelectric power, will result in a straight line with a slope E_S that is equal to E_σ (apart from the $1/q$ factor) and an intercept at $1/T = 0$ equal to kA/q as is schematically depicted in Fig. 2.10. Note that in the figure we have multiplied the Seebeck coefficient by the (positive) elementary charge e rather than by q which reflects the negative values of the thermoelectric power for conduction by electrons.

We shall go one step further and combine thermoelectric power and dc conductivity to obtain

$$Q(T) = \ln(\sigma(T)\,\Omega\,\mathrm{cm}) + \frac{q}{k}\,S(T) \tag{2.6.28}$$

which for our Standard Model gives

$$Q(T) = \ln(\sigma_0\,\Omega\,\mathrm{cm}) + A \tag{2.6.29}$$

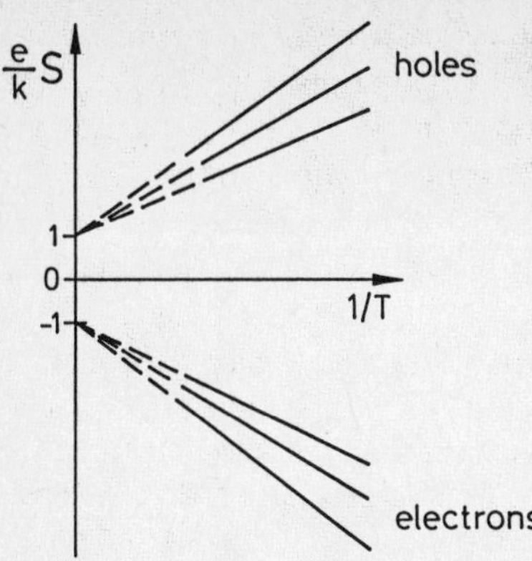

Fig. 2.10. Arrhenius plot of the thermoelectric power for conduction by electrons and by holes as predicted by the Standard Transport Model

and does no longer contain the strongly temperature dependent activation energy factors. Note that for the standard model the Q function is entirely independent of temperature.

We have treated in this subsection the cases of transport by electrons and by holes separately. In the following we shall always discuss transport by electrons unless otherwise stated. The extention to the case of holes and to ambipolar conduction will be straightforward in all cases.

2.7 Hopping Transport

In the previous sections we have always assumed that localized states in the pseudogap do not contribute to the transport at all. This cannot be true in general since electrons in localized states can tunnel into adjacent localized states. The energy difference between these states is bridged by the emission or absorption of phonons. This mode of transport is called hopping.

Let us consider first the transition probability between localized states centered around "sites" at $\boldsymbol{R}_i$ and $\boldsymbol{R}_j$ at energies ε_i and ε_j, respectively. The wavefunctions of these basis states are generally assumed to decay exponentially

$$\Psi_i(\boldsymbol{r}) \cong \exp(-|\boldsymbol{r} - \boldsymbol{R}_i|/a_{\text{loc}}) \tag{2.7.1}$$

with a common decay length, a_{loc}. These states are not orthogonal in general and, therefore, cannot be eigenfunctions of the Hamiltonian of the system. Miller and Abrahams [1960] have assumed that the proper eigenstates can be approximated as the bonding and the antibonding linear combination of pairs of states (2.7.1). This leads to $\xi_{\text{loc}} = a_{\text{loc}}$. A transition from the bonding to the antibonding state is equivalent to the transfer of most of the charge from the basis state of lower energy to the state of higher energy. In thermal equilibrium and in a single-phonon approximation this transfer rate is given by [Ambegaokar, Halperin and Langer, 1971]

$$W_{ij}^0 = \begin{cases} \Gamma_0 \exp[-2\,R_{ij}/\xi_{\text{loc}} - \beta(\varepsilon_j - \varepsilon_i)] & \varepsilon_i < \varepsilon_j \\ \Gamma_0 \exp(-2\,R_{ij}/\xi_{\text{loc}}) & \varepsilon_i > \varepsilon_j \end{cases} \tag{2.7.2}$$

where R_{ij} is the distance between the sites. The superscript "0" is used to indicate that there is no applied external field. For a real transition to take place the initial state must be occupied and the final state must be empty. For nondegenerate statistics we, therefore, obtain the equilibrium hopping rate

$$\Gamma_{ij}^0 = \Gamma_0 \exp[-2 R_{ij} / \xi_{\text{loc}} - \beta(|\varepsilon_i| + |\varepsilon_j| + |\varepsilon_i - \varepsilon_j|) / 2] \tag{2.7.3}$$

which is symmetric in i and j. The equation holds if the energies involved are large compared to $1/\beta$ and if the Fermi energy is set equal to zero. Even if multiphonon processes are considered the hopping rates are given by (2.7.3) provided the electron-phonon coupling constant is not too large [Emin, 1974; Nagel and Thomas, 1977].

We are primarily interested in the change of the transition rates due to an external field $\boldsymbol{F}$. The external field changes the transition rates and also the occupation of the states which is reflected by the changes of the local electrochemical potential. If we linearize the transition rate with respect to the applied field [Brenig et al., 1971, 1973] we obtain

$$\Gamma_{ij}(\boldsymbol{F}) - \Gamma_{ji}(\boldsymbol{F}) = e\,\beta\,\Gamma_{ij}^0\,(V_i^{\text{int}} - V_j^{\text{int}}) \tag{2.7.4}$$

where the potential V_i^{int} is the sum of the applied potential and the electrochemical potential at the site i. The system, therefore, can be represented by an equivalent electrical network of conductances

$$G_{ij} = e^2\,\beta\,\Gamma_{ij}^0 \tag{2.7.5}$$

connecting all pairs of sites. The problem of calculating the current through the sample can be replaced by the problem of finding the conductivity of the equivalent circuit [Ambegaokar, Halperin and Langer, 1971].

This task is considerably simplified by the fact that the conductances G_{ij} depend exponentially both on the spatial distance of the sites and on the site energies (note again that all energies are given with respect to E_{F}). There are a few very large conductances which, however, do not determine the network conductivity because they are in series with much smaller conductances. The numerous very small conductances, on the other hand, are also not representative for the network conductivity because they are bypassed by larger conductances.

A good estimate for the network conductivity can be obtained by the following percolation treatment [Ambegaokar, Halperin and Langer, 1971; Pollak, 1972]: we start with an "empty" network where all sites are unconnected and solder in the conductances G_{ij} in the order of their magnitude, largest first. Figure 2.11 gives a schematic picture of a small part of the network. We will obtain isolated bonds first and then, as more bonds are soldered in, small isolated clusters of sites connected by bonds. If we continue putting in smaller and smaller conductances the clusters of interconnected sites grow and merge until finally, at the "percolation threshold", the first contiguous cluster extending through the entire sample forms. This cluster is called the critical path. The conductance for

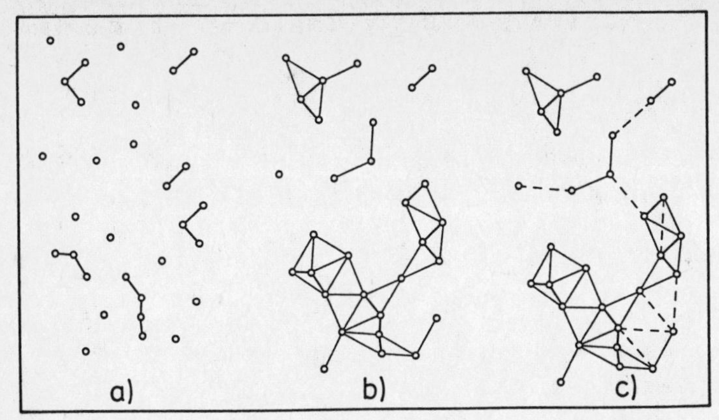

Fig. 2.11a–c. A disordered bond percolation network at different bond concentrations. (a) and (b) have isolated clusters only, (c) the first "critical" path appears

which the percolation threshold is reached is called the critical conductance and is representative for the network conductivity.

Percolation theory (see Shante and Kirkpatrick [1971] for an introduction and Kirkpatrick [1973] for a good review) considers in its simplest form a crystalline network of sites. Bonds are distributed at random between nearest neighbour sites and the formation of a critical path is monitored. For crystalline lattices we obtain the result [Ziman, 1968] that the critical bond concentration, p_c, i.e., the bond concentration at the percolation threshold, is rather independent of the lattice coordination. Pike and Seager [1974] have extended the percolation treatment to disordered lattices and find a somewhat different value for p_c as compared to crystalline lattices.

The application of percolation theory to hopping transport is straightforward: The sites of the percolation theory are identified with the sites of the localized states. The conductances G_{ij} are identified with the bonds of the percolation theory if $G_{ij} > G_c$. The value of G_c must then be determined in such a way that the bond concentration equals p_c. A detailed description of the percolation aspects of hopping transport can be found in the book of Shklovskii and Efros on heavily doped semiconductors [1984].

2.7.1 R Percolation

We consider localized states with a density n_{loc} which are in a band of width ΔE centered around $E = 0$. In the limit that $\beta\Delta E$ is small compared to the mean value of R_{ij}/ξ_{loc} for nearest neighbours the geometry of the critical path is determined entirely by the spatial distances of the sites. The bonding criterion for this "R-percolation problem" is $R_{ij} \le R_c$ with $2\,R_c = -\xi_{loc}\ln(G_c)$. A given site can, therefore, have bonds to all sites within a sphere of radius R_c. The average site therefore, has

$$p = \frac{4\pi}{3} R_c^3\, n_{loc} \qquad (2.7.6)$$

bonds. Since p must be equal to p_c in order to have a critical path we find for the critical conductance

$$G_c = \beta e^2 \exp(-2 R_c/\xi_{loc} - \beta \delta E) \tag{2.7.7}$$

where δE is of the order of the width of the band of localized states. We, therefore, expect again a conductivity of the form

$$\sigma = \sigma_0 \exp(-\beta \delta E) \quad . \tag{2.7.8}$$

If the band of localized states is centered around E_L with respect to E_F we obtain

$$\sigma = \sigma_0 \exp[-\beta(E_L + \delta E - E_F)] \quad . \tag{2.7.9}$$

The prefactor, σ_0, depends on the site density

$$\sigma_0 \sim \exp\left(-\frac{2R_c}{\xi_{loc}}\right) = \exp\left[-\frac{2}{\xi_{loc}}\left(\frac{3p_c}{4\pi n_{loc}}\right)^{1/3}\right] \tag{2.7.10}$$

and the "activation energy" exceeds $E_L - E_F$ by δE. Since in this limit transport is essentially between nearest neighbours in space this case is called nearest neighbour hopping.

It is not quite clear which energy is being transported by the hopping carrier, the energy of the initial site, that of the final site, or some mean. It has been assumed [Zvyagin, 1973; Overhof, 1975; Emin, 1975] that the mean energy is transported such that the Peltier coefficient reads

$$\Pi = \frac{1}{q}(E_L - E_F) \tag{2.7.11}$$

and the Seebeck coefficient is given by $S = \Pi/T$. Note that there is no heat of transport associated with hopping motion and that there is a finite difference, $\delta E \cong \Delta E/2$ between the "activation energy" of the thermopower and that of the dc conductivity which is a measure of the energy spread of the localized states. Hence we will have $E_Q = \delta E$ and $Q_0 = \ln(\sigma_0 \, \Omega\, \mathrm{cm})$. The latter value will depend strongly on the density of localized states, n_{loc}.

R hopping is known from doped crystalline semiconductors [Busch and Labhardt, 1946; Hung and Gleismann, 1950] where hopping in the dopant states is observed at lower temperatures for compensated samples only. This was predicted by Conwell [1956] and by Mott [1956] and experimentally verified by Fritzsche [1958, 1959, 1960]. The theory of Miller and Abrahams [1960] was in fact developed for this transport problem, see [Fritzsche and Cuevas, 1960; Pollak and Geballe, 1961].

2.7.2 Variable Range Hopping

In the case of R percolation the critical path is determined by the spatial distances of the sites. If we lower the temperature (for a fixed bandwidth ΔE) the sites with energies close to E_F will carry most of the current [Mott, 1968] and the geometry of the critical path will depend on temperature. We consider a distribution of localized states for which the density of states, $g(E)$, is independent of energy for simplicity. Mott [1969] has given an argument which shows the physical picture: We determine the average number of bonds for a given site at $E = E_F$ (which again is set equal to zero). We count all sites to be linked by a bond if the distance in space to the given site is less than R_c and if the energetic distance does not exceed ΔE. There will be on the average p bonds emanating from the given site where

$$p = \frac{4\pi}{3} R_c^3 \, g(E_F) \, 2 \, \Delta E \quad . \tag{2.7.12}$$

We equate p with p_c which then relates R_c with ΔE. We thus can write the critical conductance

$$G_c = \beta \, e^2 \, e^{-\zeta} \tag{2.7.13}$$

with

$$\zeta = \frac{2 R_c}{\xi_{loc}} + \frac{3 p_c}{4\pi \, R_c^3 \, 2 \, g(E_F) \, kT} \quad . \tag{2.7.14}$$

The minimum value for $\zeta = \zeta_c$ (i.e., the largest value for the conductance) is obtained for R_c such that

$$\left. \frac{d\zeta}{dR_c} \right|_{R_{opt}} = 0 \tag{2.7.15}$$

which results in

$$R_{opt}^4 = \frac{9 \, p_c \, \xi_{loc}}{16\pi \, g(E_F) \, kT} \quad . \tag{2.7.16}$$

Inserting this result into (2.7.14) we obtain

$$\zeta_c = \left[\frac{256}{9\pi} \, \frac{p_c}{g(E_F) \, kT \, \xi_{loc}^3} \right]^{1/4} \tag{2.7.17}$$

or

$$G_c = \beta \, e^2 \, \exp[-(T_0/T)^{1/4}] \tag{2.7.18}$$

with

$$T_0 = \frac{256}{9\pi} \, \frac{p_c}{\xi_{loc}^3 \, g(E_F) \, k} \tag{2.7.19}$$

which is Mott's famous $T^{-1/4}$–law for "Variable Range Hopping" at the Fermi energy. The conductivity will, therefore, be given by

$$\sigma = \sigma_0 \, \exp[-(T_0/T)^{1/4}] \quad . \tag{2.7.20}$$

The prefactor, σ_0, depends on the details of the wave functions and of the electron–phonon coupling [Würtz and Thomas, 1978]. We expect that it does not change if, e.g., the density of states is altered. Instead the slope, T_0, is increased by a decrease of $g(E_F)$. For a more exact treatment (which becomes mandatory if a strongly energy dependent density of states distribution $g(E)$ is considered) one has to modify this procedure: The applicability of percolation theory to hopping transport problems rests on the assumption that the distribution of bonds on the different sites is random. This approximation is fulfilled approximately for R-hopping systems. In systems where Variable Range Hopping is observed the bond distribution is of course highly correlated to the site energy. This correlation can be included [Pollak, 1972; Overhof, 1975, 1976; Overhof and Thomas, 1976] resulting in rather lengthy expressions which modify the expression for T_0 by a numerical factor of order unity. For power law densities of states

$$g(E) \sim |E^s| \tag{2.7.21}$$

these formulae result in a conductivity

$$\sigma(T) = \sigma_0 \exp[-(T_0/T)^{(s+1)/(s+4)}] \quad . \tag{2.7.22}$$

The thermoelectric power for a Variable Range Hopping system can be evaluated in a similar way [Zvyagin, 1973; Overhof, 1975]. In the limit that there is no repulsion between electrons of opposite spin at the same site the Seebeck coefficient vanishes for a density of states distribution that is symmetric with respect to E_F. In the opposite limit when the states can be singly occupied only one obtains $S = k \ln 2$ [Butcher, 1974; Chaikin and Beni, 1976]. For a density of states distribution for which the logarithmic derivative at the Fermi energy is small compared to $[kT\,(T_0/T)^{1/4}]^{-1}$ one obtains

$$S = \frac{k^2}{6e}\, p_c^2\, \sqrt{T_0 T}\, \frac{d}{dE} \ln g(E)\,|_{E=E_F} \quad . \tag{2.7.23}$$

Theye et al. [1980] show that the low temperature thermoelectric power of sputtered a-Ge can indeed be described approximately by this formula.

Hopping in tails is often considered as a possible transport mechanism. Here of course R hopping cannot be assumed but Variable Range Hopping away from the Fermi energy must be considered. An approximate treatment has been given by Hamilton [1972], by Grant and Davis [1974], and by Grünewald and Thomas [1979]. Grünewald et al. [1985] have applied the mean field theory due to Movaghar [1981] to this problem. For a density of states distribution that rises exponentially (until it finally bends over) one obtains the result that both conductivity and thermopower appear to be activated with a finite difference between the activation energies depending on the slope of the tail.

For more details we refer the reader to the extensive theoretical literature on the problem of hopping motion [Movaghar, 1981; Movaghar and Schirmacher, 1981; Movaghar et al., 1983; Böttger and Bryksin, 1976, 1985; see also Butcher, 1985].

2.7.3 Small Polaron Transport

In disordered semiconductors there is always the possibility that excess carriers form small polarons by the following mechanism: an excess electron (for holes the argument is mutatis mutandis the same) occupying a localized state polarizes its environment due to the electron-phonon coupling which increases with decreasing localization length, ξ_{loc}. The static distortion of the environment decreases the excess carrier's energy and eventually leads to a more localized state, etc. The excess electron dressed with the deformation cloud is called a small polaron [see, e.g., Holstein, 1959; Emin, 1973, 1983; Emin and Holstein, 1976]. The excess carrier is bound by the polaron binding energy to its polarization cloud and most of this energy is required if the polaron performs a hop to an adjacent site. Depending on the particular model for the electron-phonon coupling a fraction of the binding energy is transported. We expect again that the activation energies of conductivity and thermopower can differ by a certain fraction of the polaron binding energy.

It is also possible in principle that the electron-phonon coupling is sufficient [Anderson, 1972; see, e.g., Emin, 1983] to localize a significant part of the conduction band. At present there seems to be no convincing experimental evidence suggesting this mechanism to be present, not even in the chalcogenide glasses [Kastner, 1985] although Emin [1983] considers small polarons as the predominant carriers in these materials. We shall discuss this point more thoroughly in Chap. 4.

2.8 The Kubo–Greenwood Formula

It is sometimes convenient to discuss the electronic transport coefficients in terms of differential conductivities. This can be done if the Kubo formula [Kubo, 1957] is evaluated in the energy representation [Greenwood, 1958]. This is easiest if inelastic processes are of no importance for the transport. We shall derive the Kubo–Greenwood formula following Mott and Davis [1971, see also Butcher, 1985]. We start with the observation that the real part of the frequency dependent conductivity, $\sigma'(\omega)$, is related to the power dissipation density P by

$$P = \tfrac{1}{2}\,|\boldsymbol{F}|^2\,\sigma'(\omega) \tag{2.8.1}$$

if the electric field, $\boldsymbol{F}$, is given by

$$\boldsymbol{F} = \mathrm{Re}\{\boldsymbol{F}_0\,\mathrm{e}^{\mathrm{i}\omega t}\};\,. \tag{2.8.2}$$

If we choose $\boldsymbol{F}$ in the x-direction we obtain from perturbation theory with

$$H' = e\,F\,x \tag{2.8.3}$$

using Fermi's Golden rule the transition rate from state n to state m

$$W_{\mathrm{nm}} = \frac{2\pi e^2}{\hbar} \frac{|\boldsymbol{F}_0|^2}{2} |\langle \mathrm{n} | x | \mathrm{m} \rangle|^2 \times [\delta(E_\mathrm{n} - E_\mathrm{m} - \hbar\omega) + \delta(E_\mathrm{n} - E_\mathrm{m} + \hbar\omega)] \ . \tag{2.8.4}$$

The contribution of this transition to the power dissipation is given by

$$W_{\mathrm{nm}}\, f_\mathrm{F}(E_\mathrm{n})\, [1 - f_\mathrm{F}(E_\mathrm{m})]\, (E_\mathrm{m} - E_\mathrm{n}) \tag{2.8.5}$$

in lowest order of the applied field. For a system of volume Ω the real part of the conductivity thus reads

$$\sigma'(\omega) = \frac{\pi e^2}{\hbar\Omega} \sum_{\mathrm{nm}} |\langle \mathrm{n} | x | \mathrm{m} \rangle|^2 f_\mathrm{F}(E_\mathrm{n})\, [1 - f_\mathrm{F}(E_\mathrm{m})]\, (E_\mathrm{m} - E_\mathrm{n}) \times [\delta(E_\mathrm{n} - E_\mathrm{m} - \hbar\omega) + \delta(E_\mathrm{n} - E_\mathrm{m} + \hbar\omega)] \tag{2.8.6}$$

which can be simplified to be

$$\sigma'(\omega) = \frac{\pi e^2 \omega}{\hbar\Omega} \sum_{\mathrm{nm}} |\langle \mathrm{n} | x | \mathrm{m} \rangle|^2 \times [f_\mathrm{F}(E_\mathrm{m}) - f_\mathrm{F}(E_\mathrm{n})]\, \delta(E_\mathrm{n} - E_\mathrm{m} - \hbar\omega) \ . \tag{2.8.7}$$

A more rigorous derivation will be given in Chap. 5. The dc limit in this formula is not straightforward in this representation because a divergency of the sum over the matrix elements must balance the $\omega \to 0$ factor. Here we assume for simplicity that the Hamiltonian of the system can be written in terms of one-electron Hamiltonians for which

$$[x, H] = \frac{\mathrm{i}\hbar}{m} p_x \tag{2.8.8}$$

holds which allows us to rewrite the matrix elements to

$$(E_\mathrm{m} - E_\mathrm{n})\, \langle \mathrm{n} | x | \mathrm{m} \rangle = \frac{\mathrm{i}\hbar}{m} \langle \mathrm{n} | p_x | \mathrm{m} \rangle \ . \tag{2.8.9}$$

We thus obtain for the real part of the conductivity

$$\sigma'(\omega) = \frac{\pi e^2 \hbar}{\Omega m^2} \sum_{\mathrm{nm}} |\langle \mathrm{n} | p_x | \mathrm{m} \rangle|^2 \delta(E_\mathrm{n} - E_\mathrm{m} - \hbar\omega) \times \frac{f_\mathrm{F}(E_\mathrm{m}) - f_\mathrm{F}(E_\mathrm{n})}{\hbar\omega} \tag{2.8.10}$$

which in the dc limit may be written in the form

$$\sigma(T) = \int_{-\infty}^{\infty} \sigma(E) \left[-\frac{d f_\mathrm{F}(E)}{dE} \right] dE \tag{2.8.11a}$$

with the differential conductivity

$$\sigma(E) = \frac{\pi e^2 \hbar}{\Omega m^2} \sum_{\mathrm{nm}} |\langle \mathrm{n} | p_x | \mathrm{m} \rangle|\, \delta(E_\mathrm{n} - E)\, \delta(E_\mathrm{m} - E) \ . \tag{2.8.11b}$$

For the Seebeck coefficient we derive by analogy [Cutler and Mott, 1969]

$$S(T) = -\frac{k}{e} \int_{-\infty}^{\infty} \beta(E - E_{\mathrm{F}}) \frac{\sigma(E)}{\sigma(T)} \left[-\frac{d}{dE} f_{\mathrm{F}}(E) \right] dE \quad . \tag{2.8.12}$$

Hindley [1970] and Friedman [1971] have evaluated these Kubo–Greenwood formulae assuming that only the states above the mobility edge contribute to the current. For simplicity it was assumed that these states are described by (2.4.4), i.e., with random phase factors (Random Phase Model). In this approximation the differential conductivity can be cast into the form

$$\sigma(E) = kT\, g(E)\, \mu(E)\, |q| \tag{2.8.13}$$

where the mobility is itself proportional to the density of states.

The Kubo–Greenwood formula thus easily reproduces the results of our Standard Transport Model within the Random Phase approximation. It is also possible in this formalism to treat the case of two separate transport channels in parallel as, e.g., will be the case for ambipolar conduction. For nondegenerate statistics and a differential conductivity that does not depend on temperature we can rewrite the Kubo–Greenwood formulae (for electrons, e.g.) as a Laplace transform

$$\sigma(T) = \beta \int_{0}^{\infty} \sigma(E) \exp[-\beta(E - E_{\mathrm{F}})]\, dE \tag{2.8.14}$$

and

$$\frac{e}{k} S(T) = \frac{\beta^2}{\sigma(\beta)} \frac{d}{d\beta} \frac{\sigma(\beta)}{\beta} \quad . \tag{2.8.15}$$

Döhler [1979] has used the inverse Laplace transform in order to obtain explicit $\sigma(E)$-curves from experimental transport data. In contrast to our Standard Transport Model he did not obtain a differential conductivity that vanishes at some energy E_{C}. Instead his $\sigma(E)$ had a long tail into the pseudogap.

In the derivation of the Kubo–Greenwood formula we have several times made use of the fact that the electronic states are eigenstates of a one-particle Hamiltonian, i.e., that there are no inelastic processes. In general that is not true for a semiconductor and, therefore, the above formulae can only be approximately correct. If the typical energy transfer in an inelastic process is of the order of kT or less the above formulae will be approximately correct since the Fermi–Dirac distribution function does not change drastically on that energy scale. The other extreme is given by, e.g., the case of Variable Range Hopping transport where the exchange of energy between electrons and phonons is the basic transport process. In the limit of a high density of states where the average $R_{ij}/\xi_{\mathrm{loc}}$ (and, therefore $\Delta E/kT$) is of order unity, even Variable Range Hopping transport can be described by the above formula (2.8.14), but certainly not by (2.8.15), because the derivation of the latter formula requires that $\sigma(E)$ does not depend on temperature.

2.8.1 The Model of Two Separate Conduction Channels

From the first observed drift mobility data [LeComber and Spear, 1970] on hydrogenated silicon it was concluded that transport at elevated temperatures is due to the extended states while at lower temperatures transport is by hopping in the tail states. If both channels are supposed to be independent we see from the Kubo–Greenwood formula (2.8.11a) that the corresponding conductivities of both channels must be added [LeComber, Madan and Spear, 1972]:

$$\sigma = \sigma_{\text{ext}} + \sigma_{\text{hopp}} \tag{2.8.16}$$

while the contributions to the thermoelectric power have to be weighted according to

$$S = \frac{\sigma_{\text{ext}}}{\sigma} S_{\text{ext}} + \frac{\sigma_{\text{hopp}}}{\sigma} S_{\text{hopp}} \tag{2.8.17}$$

(see e.g., the discussion in [Mott and Davis, 1979]). This model results in a Q function that in a limited temperature range appears to be activated. For more elevated and for lower temperatures the model predicts that the Q function has a smaller slope than in the intermediate range. The model has been used extensively by many groups in order to parametrize the transport properties. Beyer and Overhof [1984] give several examples for the Q function which follows from this model if the parameters derived from the experimental data are inserted.

3. Review of Transport Experiments

While the Variable Range Hopping transport first observed in evaporated germanium by Clark [1967], by Walley and Jonscher [1968] and later also in silicon [Walley, 1968] attracted the attention of many theorists and experimentalists in the early seventies the transport in extended states received much less attention at that time. This latter transport mechanism is predominant in hydrogenated amorphous Silicon (a-Si:H) prepared, e.g., by the glow-discharge decomposition of silane [Sterling and Swann, 1965; Chittick et al., 1969]. The picture changed entirely when Spear and LeComber [1975] proved that hydrogenated material can be effectively doped which opened the door to many industrial applications. In the time that followed the number of papers on transport experiments with a-Si:H has grown enormously. It is, therefore, impossible to mention or even to quote all the relevant experimental literature (for a review see Beyer and Overhof [1984], and Fuhs [1984] on material prepared by glow-discharge decomposition, and Anderson and Paul [1981, 1982], and Paul and Anderson [1981] for sputtered a-Si:H). Instead, we shall try so summarize that part of the experimental literature that seems to converge towards a unified picture.

It is clear that in a field which is still far from being mature and settled there are often contradictory experimental findings reported by the different research groups. It turns out, however, that many experiments performed in different laboratories yield quite comparable results. This is in particular true for transport experiments performed on moderately or heavily doped samples prepared under optimized conditions. Here we observe that the experimental results agree perfectly in many cases — the interpretations given, however, are usually more controversial.

For doped a-Si:H and a-Ge:H prepared by the glow-discharge deposition the doping level usually is given by a specification of the doping gas content in the silane or germane mixture prior to decomposition. We shall adopt this notation although we note that this nominal doping level may not be a good measure for the density of electrically active dopants in the samples.

3.1 dc Conductivity

The measurement of the temperature dependence of the dc conductivity is performed routinely in most laboratories in order just to characterize the samples. Because the samples can be prepared as thin films only all transport measurements are plagued by the problems inherently connected with thin films:

i) The conductivity can be influenced by surface contamination [Tanielian et al., 1978] and by interface states [Solomon et al., 1978; Solomon, 1981]. These may lead to depletion layers that reduce the effective sample thickness or to accumulation layers of high conductivity to the extent that the sample conductance is determined entirely by the accumulation layer.
ii) The conductivity of the film depends on the film thickness [Ast and Brodsky, 1979, 1980] in particular for very thin films. This effect can be correlated with the observation [Müller et al., 1980; Beyer and Wagner, 1981] that the hydrogen content of thick films changes near surfaces and interfaces.
iii) The samples are sometimes porous. From hydrogen effusion studies it is known [Beyer and Wagner, 1982] that in boron-doped samples the hydrogen migration is no longer limited by diffusion, a strong indication of porosity.

Besides these problems there are other difficulties related with contacts etc., that are common in the semiconductor field. Practically all reported measurements have been performed in the standard gap cell configuration shown schematically in Fig. 2.7. Very little information has been reported on measurements with four-probe or sandwhich-type arrangements.

3.1.1 Transport in Extended States

Figure 3.1 shows an Arrhenius plot of the conductivity for n-type a-Si : H doped by phosphorus. Figure 3.2 shows corresponding data for p-type samples doped by boron (undoped a-Si : H and a-Ge : H samples are found to be always n-type).

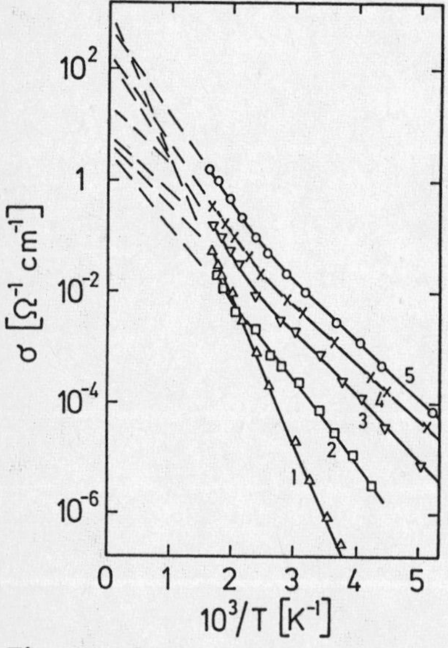

Fig. 3.1. Conductivity for n-type a-Si : H films as a function of inverse temperature. $[PH_3]/[SiH_4]$ = 10^{-6} (*1*), 3×10^{-6} (*2*), 2.5×10^{-4} (*3*), 10^{-3} (*4*), and 10^{-2} (*5*) (after [Beyer et al., 1977])

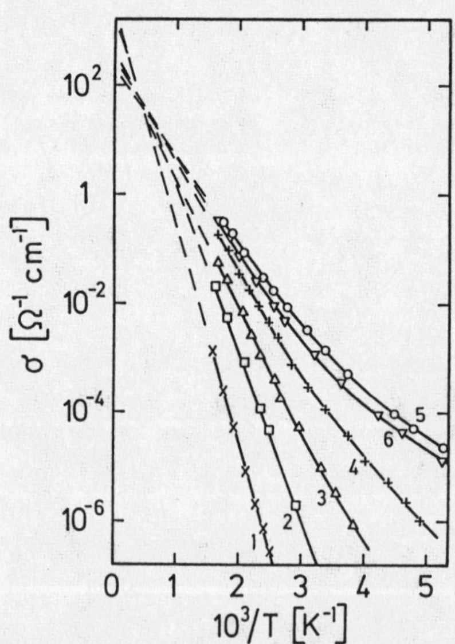

Fig. 3.2. Conductivity for p-type a-Si : H films as a function of inverse temperature. $[B_2H_6]/[SiH_4] = 10^{-5}$ (*1*), 10^{-4} (*2*), 10^{-3} (*3*), 10^{-2} (*4*), 5×10^{-2} (*5*), and 2×10^{-1} (*6*) (after [Beyer and Mell, 1977])

Results for doping by phosphorus agree qualitatively with those obtained for doping by As, Li, K, and Na [Beyer, Fischer and Overhof, 1979; Beyer et al., 1979, 1980; Jan et al., 1979], and also with those doped by Ga [Gross et al., 1987]. Similar data have been obtained by several other groups. In particular for moderately and highly doped samples the results obtained in different laboratories agree perfectly.

For n-type samples a few ppm of phosphine are sufficient to raise the low temperature conductivity by a few orders of magnitude. With increasing doping the rate of this increase considerably slows down until, at the very extreme doping levels (which better should be termed alloying), the conductivity is again slightly decreased. Similar results are also obtained for sputtered hydrogenated silicon [Anderson and Paul, 1980, 1981].

In contrast to our expectation from the Standard Transport Model (see Fig. 2.8) we see that the conductivity data do not appear to be strictly activated: for n-type samples there are marked "kinks" in either direction [Beyer, Medeišis, and Mell, 1977] while for p-type material there appears a continuous change of slope, most prominent for the highly doped samples.

Because of this deviation from the expected activated behaviour there is no obvious procedure to determine the transport parameters σ_0 and E_σ. The most popular method is the following: one ignores the kinks and performes a linear force-fit to the experimental data from a limited temperature range, in most cases for temperatures around room temperature. The intercept at $1/T = 0$ is then identified with the apparent prefactor of the conductivity, σ_0^*, while the slope is taken as the activation energy, E_σ^*. Figure 3.3 shows that the resulting transport parameters observe a strange law [Rehm et al., 1977; Carlson and Wronski, 1979; Spear et al., 1980; Fritzsche, 1980]

$$\ln(\sigma_0^* \, \Omega \, \mathrm{cm}) = \mathrm{const.} + E_\sigma^*/E_{\mathrm{MNR}} \tag{3.1.1}$$

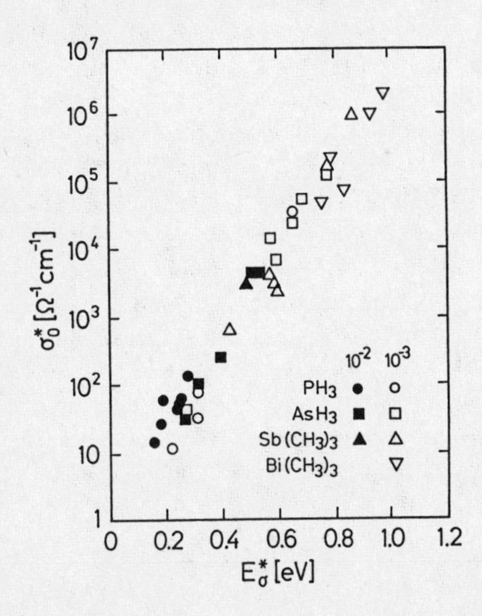

Fig. 3.3. The Meyer–Neldel rule for a-Si:H doped with various dopants (from [Carlson and Wronski, 1979])

which is called the Meyer–Neldel rule (MNR) since Meyer and Neldel [1937] found this rule in their work on the transport properties of baked semiconductor powders. Figure 3.3 emphasizes the enormous spread of prefactors, σ_0^*, that are found for doped samples. For undoped samples the spread is somewhat smaller but still exceeds 3 orders of magnitude [Spear et al., 1980].

It is clear that such a large variation of the observed prefactors cannot be explained by a corresponding spread of the quantities that enter into the calculation of the microscopic prefactor (see (2.6.14) and (2.6.16), respectively). Note also that for systems where the Meyer–Neldel rule, i.e., the simultaneous variation of σ_0^* and E_σ^*, is valid the spread of the actual conductivity values at elevated temperatures is by several orders of magnitude smaller than the spread in prefactors. In fact, if the MNR would hold for all temperatures (3.1.1) predicts that the different conductivity curves would intersect at a temperature

$$T^{\mathrm{i}} = E_{\mathrm{MNR}}/k \ . \tag{3.1.2}$$

We shall show in Chap. 7 that the slopes, E_σ^*, and the intercepts, σ_0^*, do not have the physical meaning of prefactors and activation energies as in (2.6.13) and (2.6.15). We, therefore, have denoted by an asterisk (and shall continue to do so) all quantities that are derived from an Arrhenius plot of experimental data. We shall also use the epitheton "apparent" to warn against a naïve interpretation of these parameters.

3.1.2 The Staebler–Wronski Effect and Metastability

Staebler and Wronski [1977, 1980] have shown that upon heavy illumination the dc conductivity of a-Si:H samples is drastically decreased. This Staebler–Wronski effect (SWE) has been widely studied since the change in conductance (accompanied by a change in photoconductivity) leads to a deterioration of the performance of solar cells. The sample can be brought back from the light soaked state (called state B) to its original state (called state A) by annealing. Similar to the control of the conductivity by doping this reversible effect is characterized by a simultaneous variation of σ_0^* and E_0^* leading to another MNR [Wagner et al., 1983; Irsigler et al., 1983] which can be derived for every sample separately. Figure 3.4 shows a compilation of the MNR data obtained for several n-type samples. Note that in this figure the σ_0^* values seem to saturate at a value of $10^4\ \Omega^{-1}\,\mathrm{cm}^{-1}$. For the part of the figure where (3.1.1) is observed we obtain $E_{\mathrm{MNR}} = 0.043$ eV as opposed to $E_{\mathrm{MNR}} = 0.067$ eV derived from Fig. 3.3.

As amorphous semiconductors are not in thermal equilibrium but at most in a metastable state one should observe that the transport data are affected by the thermal history of the sample. In fact transport experiments have always been plagued by nonreproducible results and hysteresis effects [Ast and Brodsky, 1979]. The standard procedure to cope with this problem is to anneal the samples at elevated temperatures (in order to get rid of the SWE) and to perform the transport measurements upon subsequent slow cooling of the samples. This procedure in fact guarantees reproducible results in most cases.

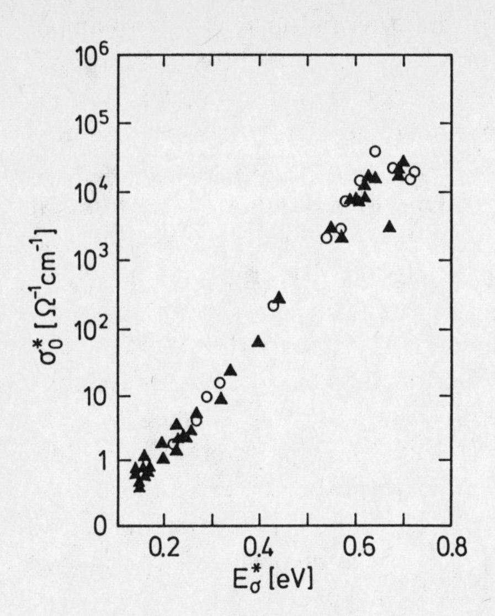

Fig. 3.4. Meyer–Neldel rule for light soaked n-type a-Si : H at different annealing stages (from [Irsigler et al., 1983])

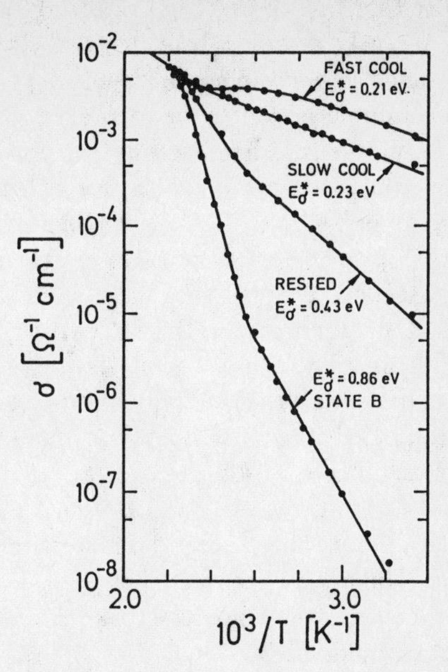

Fig. 3.5. Conductivity versus inverse temperature for n-type a-Si : H measured immediately after annealing for different cooling rates, after storing in the dark for 7 months, and after illumination (from [Kakalios and Street, 1986])

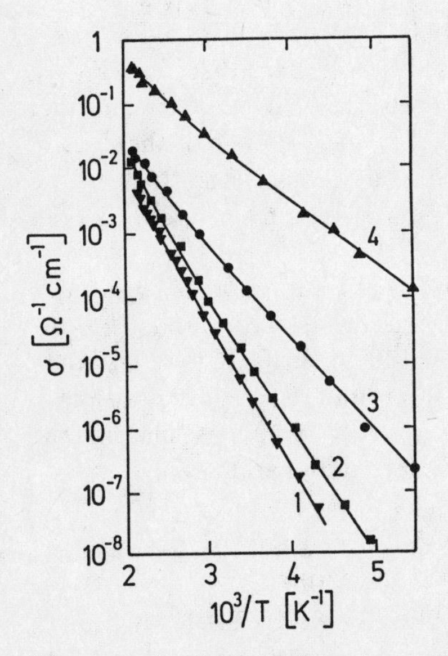

Fig. 3.6. Conductivity for n-type a-Ge : H films as a function of inverse temperature. $[PH_3]/[GeH_4] = 0$ (*1*), 10^{-4} (*2*), 10^{-3} (*3*), and 2×10^{-2} (*4*) (after [Hauschildt, Stutzmann et al., 1982])

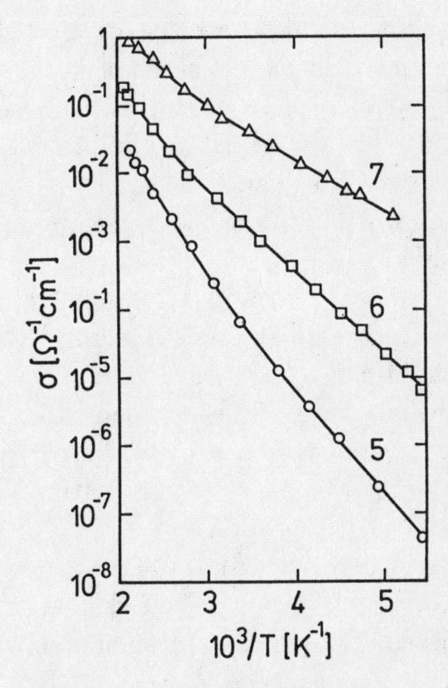

Fig. 3.7. Conductivity for p-type a-Ge : H films as a function of inverse temperature. $[B_2H_6]/[GeH_4] = 10^{-3}$ (*5*), 5×10^{-3} (*6*), and 3×10^{-2} (*7*) (after [Hauschildt, Stutzmann et al., 1982])

Recently however Kakalios and Street [1986] and Street et al. [1986, 1987] have shown that the conductivity depends on the thermal history of the sample if the sample is cooled more rapidly. Figure 3.5 shows conductivity data for a single sample that has been subject to different heat treatments and illumination cycles. The figure illustrates the extent to which the conductivity of a single sample can be altered in a reversible cycle at lower temperatures. It also serves to demonstrate our reservations about the physical meaning of slopes and intercepts: The steepest part of the state B conductivity curve (which of course can be measured only irreversibly, i.e., with rising temperature) extrapolates to $E_\sigma^* = 2\,\mathrm{eV}$ with $\sigma_0^* = 10^{18}\,\Omega^{-1}\,\mathrm{cm}^{-1}$!

Compared with the large number of papers on the transport properties of a-Si : H much less attention has been paid to a-Ge : H [Hauschildt, 1982; Hauschildt, Stutzmann et al., 1982; Jones et al., 1976, 1979]: Figures 3.6 and 3.7 show conductivity data for P-doped and B-doped a-Ge : H samples, respectively [Hauschildt, Stutzmann et al., 1982]. The results are in many respects similar to those of a-Si : H. The n-type samples show kinks in the Arrhenius plot (which are not quite pronounced in Fig. 3.6) whereas for the p-type samples we observe a continuous change of slope instead.

3.1.3 Transport in Localized States

For unhydrogenated silicon and germanium one observes at lower temperatures a dc conductivity that is described as Variable Range Hopping [Walley, 1968; see e.g., Beyer and Stuke, 1975; Beyer et al., 1975] with a value for T_0 that is about 10^8 K and rather insensitive to changes of the density of dangling bond states. From ESR experiments on comparable samples [Stuke, 1976] it is known that the spin density can be varied in these materials by more than one order of magnitude. In contrast to the prediction of (2.7.19) the change of the density of states does not lead to a significant change of the T_0 value also. Instead the σ_0 value (which should not be strongly affected) is altered by several orders of magnitude.

It is possible to simulate such an effect by the assumption of a suitable density of states distribution around the Fermi energy [Overhof and Thomas, 1976; see also Overhof, 1976; Ortuno and Pollak, 1983]. In fact, the density of states distribution can be chosen similar to that produced by a distribution of dangling bonds with a positive correlation energy of about 0.3 eV pinning the Fermi energy. The degree of insensitivity of the observed T_0 parameters, however, is not easily understood.

For hydrogenated material a contribution of Variable Range Hopping to the conductivity is observed if the defect density is made artificially high. This can be obtained by, e.g., ion implantation or by annealing at temperatures where most of the hydrogen is driven out [Hauschildt, 1982; Hauschildt, Fuhs, et al., 1982]. Little work has been done to determine the hopping transport parameters in these materials.

3.2 Peltier Effect and Thermoelectric Power

Direct quantitative measurements of the Peltier coefficient are extremely rare. There is an ingenious experiment by Dersch and Amer [1984] where the heat production could be measured by optical beam deflection spectroscopy as a function of one space coordinate: a current of frequency ω ($5 - 10$ Hz) produces Peltier heat of the same frequency but Joule heat of frequency 2ω. While the Peltier heat is produced at the electrodes only, the Joule heat should be generated uniformly over the sample. Dersch and Amer show that for samples with good ohmic contacts this is indeed the case.

There are relatively few data on the thermoelectric power [Grigorovici et al., 1967; Jones et al., 1977; Beyer and Mell, 1977; Beyer et al., 1977; Anderson and Paul, 1981, to mention a few]. Figures 3.8 and 3.9 show examples for the Seebeck coefficient in an Arrhenius plot for n-type and p-type a-Si : H, respectively. If we compare these data with the conductivity data shown in Figs. 3.1 and 3.2 for the same samples we see that the kinks and the continuous change of slope are present in both sets of data in a very similar way. It should be noted that the extrapolated intercepts of the thermopower at $1/T = 0$ are much too large to be identified with the heat of transport term of (2.6.27). This discrepancy is even more pronounced if the extrapolation is made from the low temperature data.

Closer inspection of the data shows that at any temperature the local slope of the conductivity exceeds that of the thermoelectric power by an amount that varies for different samples. This was already observed for evaporated amorphous silicon and germanium samples for which the high temperature transport data could be interpreted in terms of transport in extended states [Beyer and Stuke, 1974]. The kinks observed in the Arrhenius plots for the conductivity and the

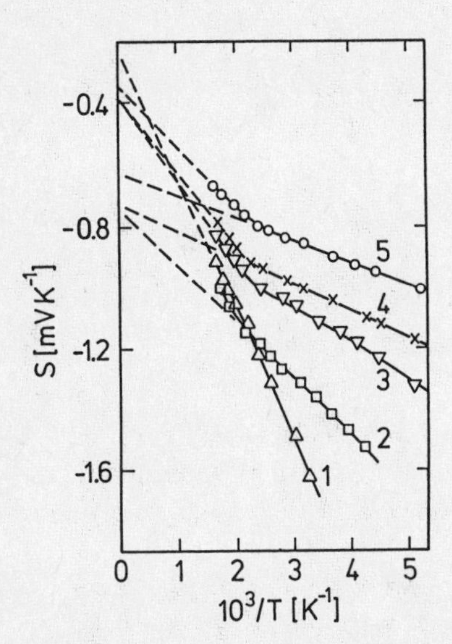

Fig. 3.8. Thermoelectric power for the n-type a-Si : H samples of Fig. 3.1

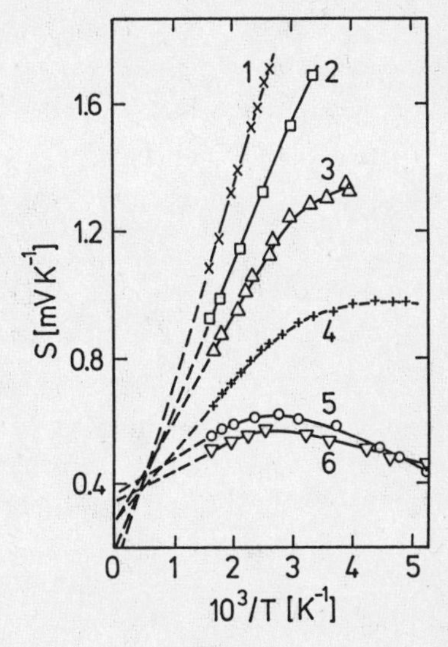

Fig. 3.9. Thermoelectric power for the p-type a-Si : H samples of Fig. 3.2

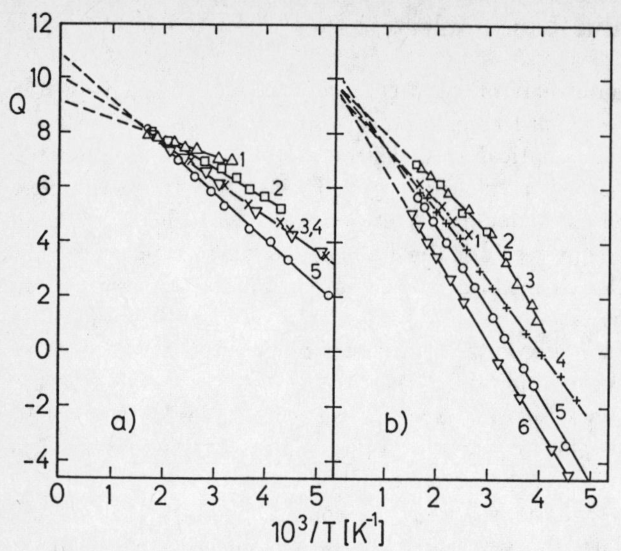

Fig. 3.10. Q as a function of inverse temperature for **(a)** n-type a-Si : H and **(b)** p-type a-Si : H from the combination of the data of Figs. 3.1 with 3.8 and of Figs. 3.2 with 3.9, respectively

thermoelectric power disappear entirely if we combine the data to the quantity $Q(T)$ [Beyer and Overhof, 1979]. Figure 3.10 shows experimental data for Q in an Arrhenius plot both for n-type and for p-type material. Within experimental accuracy the data follow the empirical relation

$$Q(T) = Q_0^* - \beta E_Q^* \tag{3.2.1}$$

with a value of $Q_0^* = 10 \pm 1$ and with E_Q^* in the $0.05\,\mathrm{eV} < E_Q^* < 0.25\,\mathrm{eV}$ range in all but the very extreme cases [Overhof and Beyer, 1983; see also Beyer and Overhof, 1984]. For undoped a-Si : H the smaller values of E_Q^* are observed, for highly n-doped samples the E_Q^* values usually do not exceed 0.15 eV, while p-type samples generally have the highest values of E_Q^*. For all samples the value of E_Q^* can be increased beyond 0.2 eV by

i) annealing at temperatures that are sufficient to drive out a good deal of the hydrogen content [Hauschildt, Fuhs et al., 1982].
ii) extensive illumination: the Staebler–Wronski effect is accompanied by a reversible increase of E_Q^* [Hauschildt, Fuhs et. al., 1982].

From Fig. 3.10a one might suspect that a Meyer–Neldel type rule holds for Q_0^* and E_Q^* as was reported in the first paper on the Q function [Beyer and Overhof, 1979]. In the meantime, however, it was shown that there is no systematic correlation of both quantities.

There is, however, a systematic difference between the Q data and virtually all other transport data. We have already seen that we obtain Meyer–Neldel rules for the dc conductivity and for the thermoelectric power if we change the respective "activation energies" by doping, by heat treatment, or by light exposure. We shall see in the following sections that we again obtain such rules in the space charge limited current experiment and in the field effect where we also alter the respective "activation energies" reversibly with the aid of some applied external field. For the Q functions we find on the contrary that the prefactor, Q_0^*, does not depend on the slope, E_Q^*, which, however, is increased by, e.g., increasing dopant content.

We thus see that we have Meyer–Neldel rules for all quantities that contain the $E_C - E_F$ "activation energy" factor. This strongly points towards the suspicion that the Meyer–Neldel rules might be essentially due to a temperature dependence of this factor. We shall see in Sect. 7.1 that in fact the temperature dependence of the Fermi energy causes the Meyer–Neldel behaviour in all cases.

For heavily doped and partly compensated a-Si:H Beyer and Mell [1981] have shown that the slope E_Q^* depends primarily on the boron content of the samples. We show in Fig. 3.11a data where the boron content is kept fixed and the phosphorous content is varied. As a consequence the Q data apparently fall onto a single line and no difference between n-type and p-type samples is observed in the Q function. In contrast, the data in Fig. 3.11b where the boron content is varied show a systematic increase of E_Q^*. Note that in all cases $Q_0^* = 10.5$.

In Fig. 3.12 we show conductivity and thermoelectric power data (the latter multiplied by eT, and not by qT in order to discriminate between n-type and p-type samples) for a series of compensated samples [Beyer et al., 1981]. The most interesting feature of these data is the fact that for one sample the compensation is perfect in the entire temperature range accessible to the experiment: the thermoelectric power is exactly zero. From the Kubo–Greenwood formulae (2.8.11) and (2.8.12) it is clear that in this case the contributions from electrons and holes both to the current and to the Peltier heat current must have exactly the same absolute value. The parameters of the conduction paths for electrons

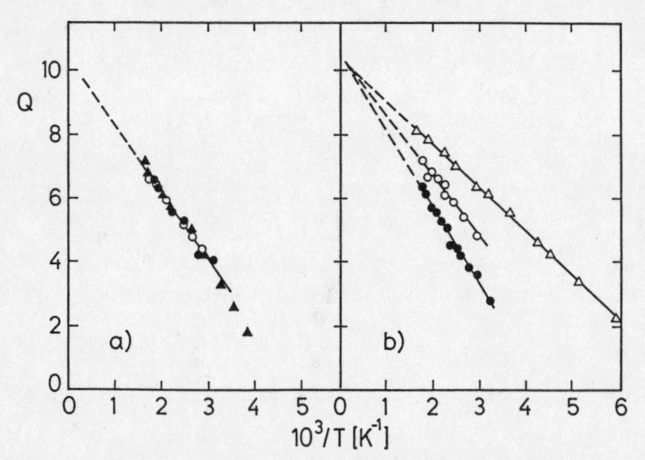

Fig. 3.11. Q as a function of inverse temperature for compensated a-Si:H (a) $[B_2H_6] = 10^{-3}$: $[PH_3] = 0$ (▲), 2×10^{-4} (•), 3×10^{-3} (○) (b) $[PH_3] = 10^{-3}$: $[B_2H_6] = 10^{-3}$ (△), 2×10^{-4} (○), 3×10^{-3} (•) full symbols are for n-type samples, open symbols for p-type samples (from [Beyer et al., 1981])

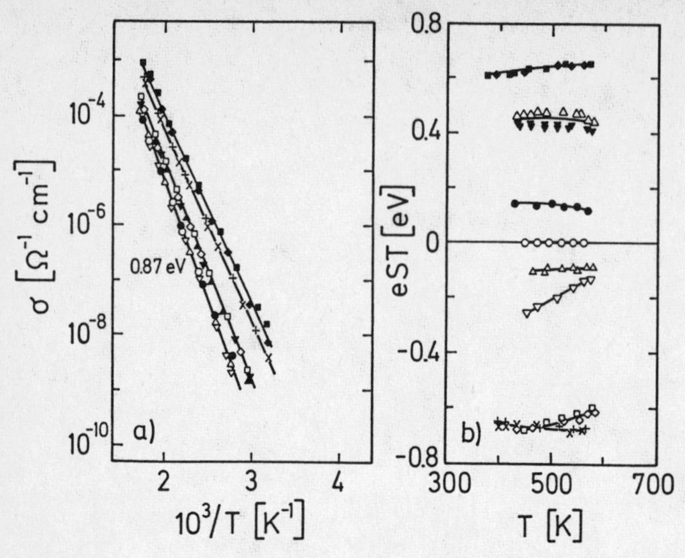

Fig. 3.12. Conductivity and thermoelectric power for a series of compensated films (from [Beyer et al., 1981])

and holes must, therefore, be quite symmetrical. We shall in Sect. 8.1.2 come back to this particular sample where we exploit the fact that E_σ^* is 0.87 eV while the prefactor, σ_0^*, equals 4700 $\Omega^{-1}\,\mathrm{cm}^{-1}$.

Inspection of the thermoelectric power data reported in the literature [see, e.g., Ghiassy et al., 1985; Anderson and Paul, 1982] confirm that in all cases except for those where a contribution of hopping conduction must be suspected the Q function is extremely well represented by (3.2.1) with $Q_0^* \simeq 10$.

For doped a-Ge : H we show in Fig. 3.13 thermoelectric power data [Hauschildt, 1982; Hauschildt, Stutzmann et al., 1982]. For heavily doped samples the Seebeck coefficient is virtually independent of the temperature but by no means small, quite the opposite to what one would expect for activated transport. If we look at the corresponding Q data displayed in Fig. 3.14 we find essentially the same picture as in the case of a-Si : H. There is a significant difference compared to the silicon data, however: For germanium the slope $E_Q^* \cong 0.15$ eV is practically independent of the doping level but the intercept, Q_0^*, for p-type samples is systematically larger than its correponding value for n-type samples. Such a difference has not been observed in a-Si : H.

We note in passing that for most chalcogenide glasses we have a Q function which is similar to that of highly doped a-Si : H with $Q_0^* \cong 10$ and $E_Q^* \cong 0.2$ eV, however, with a somewhat larger scatter in the Q_0^* data [Overhof and Beyer, 1984]. The same holds for the alloys of the a-Si : C : H and the a-Si : Ge : H systems [Hauschildt, 1982; Bort, 1986]. It seems that the very small values of E_Q^* are found exclusively in undoped a-Si : H prepared under optimized conditions while any sort of additional disorder due to doping etc., leads to larger values of E_Q^*.

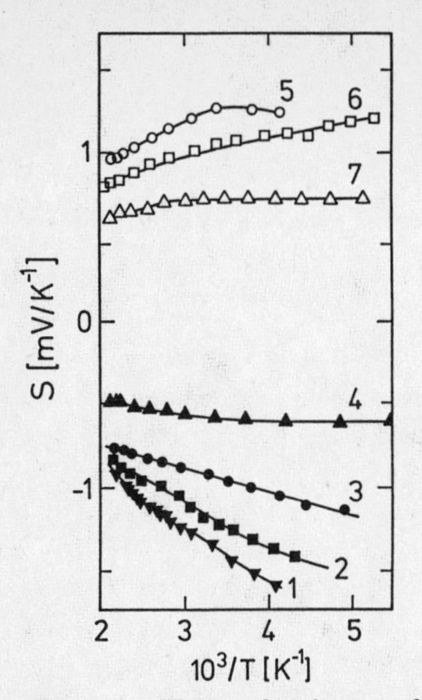

Fig. 3.13. Thermoelectric power for n-type (*full symbols*) and p-type (*open symbols*) a-Ge:H. The key is as in Figs. 3.6 and 3.7 (after [Hauschildt, Stutzmann, et al., 1982])

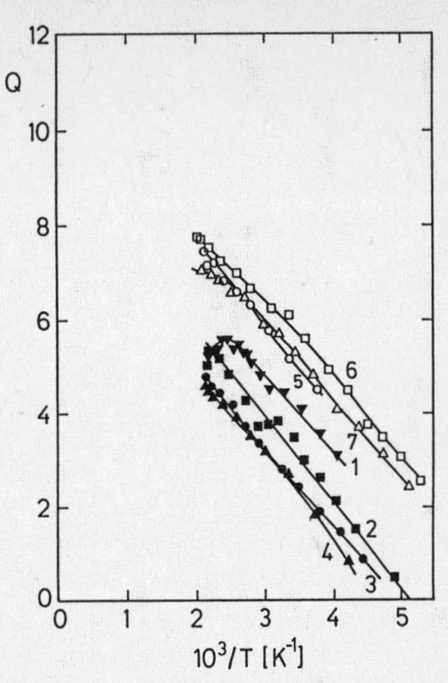

Fig. 3.14. Q for n-type (*full symbols*) and p-type (*open symbols*). The key is as in Figs. 3.6 and 3.7 (after [Hauschildt, Stutzmann, et al., 1982])

3.3 Hall Mobility

The Hall effect has been proven to be extremely useful in the investigation of crystalline semiconductors because it provides in many cases a direct and reliable measure of the density and the sign of the charge of the carriers. The experimental setup is sketched in Fig. 3.15: the current driven by the electric field $\boldsymbol{F}$ is subject to the Lorentz force caused by a magnetic Field $\boldsymbol{B}$ which is perpendicular to the plane of the sample. The Lorentz force leads to a small surface charge. This charge in turn generates the electrostatic Hall field perpendicular to the current direction which finally balances the Lorentz force on the average. For free carriers with an isotropic effective mass m^* and a constant relaxation time we find that the Hall field $\boldsymbol{F}_{\mathrm{H}}$ is given by

$$q\,\boldsymbol{F}_{\mathrm{H}} = q\,\boldsymbol{v} \times \boldsymbol{B} = |q|\,\mu_{\mathrm{H}}\,\boldsymbol{F} \times \boldsymbol{B} \tag{3.3.1}$$

which defines the Hall mobility, μ_{H}. For a constant relaxation time the Hall mobility equals the mobility determined from the dc current. The sign of the predominant carriers gives the sign of the Hall field and from the mean velocity $\boldsymbol{v}$ one easily determines (in combination with the current density) the carrier density.

For amorphous semiconductors the Hall effect cannot be interpreted easily in contrast to crystalline semiconductors because [Beyer and Mell, 1977; LeComber

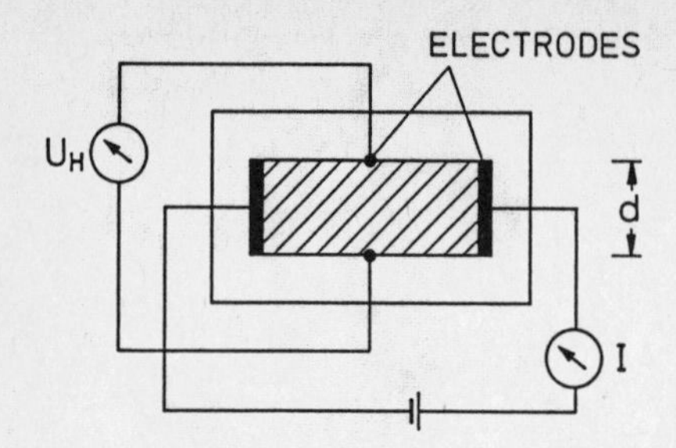

Fig. 3.15. Experimental setup for the determination of the Hall mobility. A magnetic field is perpendicular to the plane of the sample

et al., 1977; Roilos, 1978] the sign of the predominant carriers as derived from the Hall effect is opposite to that expected from all the other experiments (thermoelectric power, the doping state of the sample, and field effect). This is in fact not too astonishing as the simple-minded picture of a free particle subject to a Lorentz force cannot hold in amorphous semiconductors where the estimated mean free path of the carriers is less than the interatomic distance [Ioffe–Regel rule, 1960]. The only microscopic theories that predict a sign reversal of the Hall effect assume transport by small polarons. Emin [1977] calculates a sign reversal for small polarons moving in a structure where odd-membered rings predominate. Grünewald et al. [1981] have shown that for a more realistic model of localized states consisting of a mixture of s-type and p-type orbitals the sign reversal for the Hall effect can as well be obtained for a movement in an even-membered ring structure. For transport in extended states there is to date no satisfactory explanation for the sign anomaly of the Hall effect.

If we derive a Hall mobility from the experimental data using the Standard Transport Model of Sect. 2.6 and ignore the sign anomaly we find results as shown in Fig. 3.16 for n-type samples [Beyer et al., 1977] and for p-type samples [Beyer and Mell, 1977] of a-Si : H. We observe an approximately activated Hall mobility (with an activation energy $E^*_{\mathrm{H}} \cong (1/3)E^*_Q$) which bends over as kT becomes comparable to E^*_{H}.

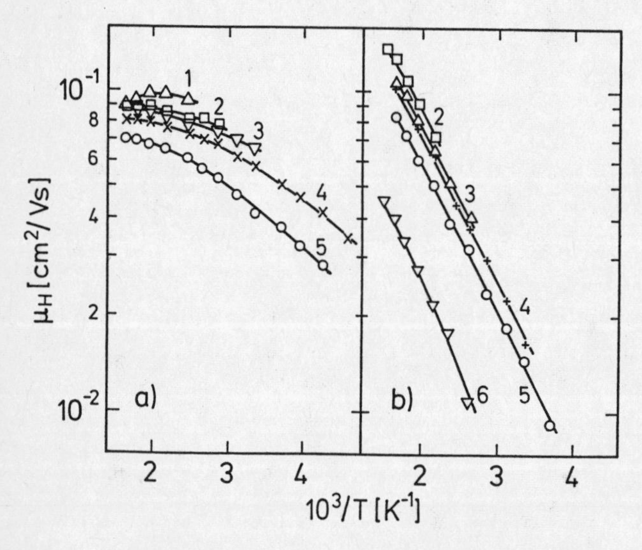

Fig. 3.16. Hall mobility for a-Si : H samples as a function of inverse temperature (**a**) n-type samples of Fig. 3.1 (from [Beyer et al., 1977]) (**b**) p-type samples of Fig. 3.2 (from [Beyer and Mell, 1977])

For highly doped n-type and p-type samples a rather strong magnetoresistance effect has been reported [Weller et al., 1981]. The only theory that explains this effect satisfactorily has been derived by Movaghar and Schweitzer [1978] for hopping transport. We, therefore, must keep in mind that at least for the highly doped samples there can be a contribution to the electronic transport by hopping in tails although it is usually not detected directly in the conductivity or thermoelectric power.

3.4 Space Charge Limited Currents

In the study of the transport properties of semiconductors one is usually interested in linear current-voltage ($I-V$) characteristics. Much care has been taken to make sure that the contacts are not blocking (which usually is believed to be true if Ohm's law holds). The idea of a dc transport experiment is to monitor the response of the mean carrier velocity to the applied external voltage without altering the mean carrier density. This is in fact how the linear response theory describes Ohmic transport.

If the external voltage is increased and if the electrodes do not block the flow of the carriers additional charges will be swept into the sample. Let us assume for simplicity that we have contacts that inject additional carriers of one polarity, electrons e.g, only. The excess charges form a space charge that limits the injected charge and, therefore, the resulting current is called space charge limited current (SCLC). We shall discuss here the oversimplified case in which the space charge is distributed homogeneously. In their excellent textbook Lampert and Mark [1970] show that in most cases this simplification leads to surprisingly small errors. For a homogeneous density, n, of charge, q, in a sample of electrode distance, l, with a dielectric constant, ε_s, we have owing to the laws of electrostatics

$$nq = \frac{\varepsilon_0 \, \varepsilon_s \, V}{l^2} \tag{3.4.1}$$

as is the case for a capacitor charged with a voltage, V.

For a crystal without states in the gap we, therefore, will have a density of carriers that equals the equilibrium density plus a component that is proportional to the applied voltage, V. The resulting current will, therefore, have two components, one that is proportional to the voltage and a second one that is proportional to the voltage squared. This latter component will dominate at the higher voltage levels and lead to the typical law for space charge limited currents in crystals (in the absence of traps)

$$I \sim V^2 \ . \tag{3.4.2}$$

In an amorphous semiconductor the space charge will not primarily consist of mobile carriers. Instead, the majority of carriers will condense into states around the Fermi level. For a homogeneously charged system we, therefore, have a modified distribution function which equals the familiar Fermi–Dirac distribution function, however with the Fermi energy replaced by the quasi-Fermi energy, E_F^q.

At sufficiently low temperatures we can replace the Fermi–Dirac function by a step function if the density of states distribution around the Fermi energy is sufficiently smooth. We can in this case obtain the total density of injected electrons from the integrated density of states distribution [Orton and Powell, 1984]

$$n = 2 \int_{E_F}^{E_F^q} g(E)\, dE \tag{3.4.3}$$

(the factor 2 comes from spin degeneracy). The total carrier density is given by (see (2.6.6) and (2.6.12))

$$n = N_C \exp[-\beta(E_C - E_F^q)] \ . \tag{3.4.4}$$

In the simplest case where the density of states can be replaced by a constant we find that the current rises nearly exponentially with the applied voltage

$$I \sim V \exp(\beta\, u\, V) \tag{3.4.5}$$

with

$$u = \frac{\varepsilon_0\, \varepsilon_s}{2\, e\, g(E_F)\, l^2} \ . \tag{3.4.6}$$

From these formulae one expects that at least the density of states at the Fermi energy is most easily determined by the SCLC experiment. A major advantage of this experiment is the fact that it samples the density of states in the entire sample and is, therefore, not sensitive to surface or interface states.

It is somtimes difficult to discriminate the SCLC effect from other effects that may lead to a nonlinear $I - V$ characteristic as, e.g., will be the case if the electrodes form Schottky barriers. In this case it is somtimes possible to identify the SCLC effect by a particular scaling law. If the electronic properties of the sample remain unchanged with the alteration of the sample thickness one can show [see Lampert and Mark, 1970] that under quite general conditions the current is given by

$$I/l = f(V/l^2) \tag{3.4.7}$$

where the function $f(x)$ depends on the properties of the material but *not* on the sample thickness (note that the scaling law includes Ohmic currents as a trivial case).

The space charge limited current experiment has been performed with a-Si : H samples by several groups [DenBoer, 1981; Mackenzie et al., 1982; Solomon et al., 1984] primarily because it offers the possibility to determine directly the density of states distribution near the equilibrium Fermi energy. Figure 3.17 shows a typical plot of current-voltage curves obtained at different temperatures for a n-type a-Si : H sample. For a fixed applied voltage the relation for the current as a function of temperature is again activated

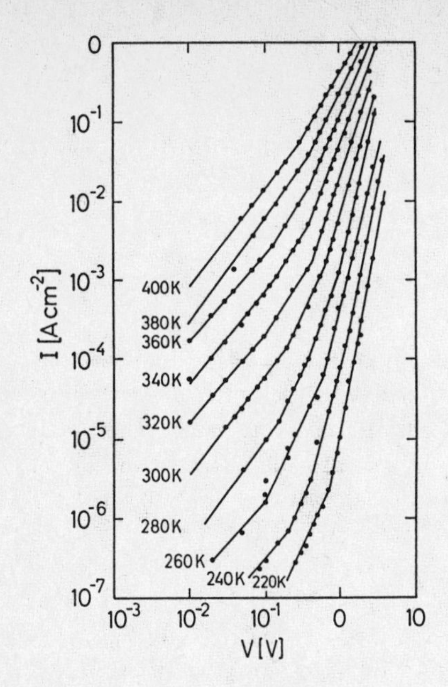

Fig. 3.17. $I-V$ characteristic for an n-type a-Si : H sample at different temperatures (from [Mackenzie et al., 1982])

$$I = I_0^* \exp(-\beta E_a^*) \tag{3.4.8}$$

with a prefactor I_0^* that depends on the activation energy E_a^* via

$$I_0^* = I_{00} \exp(E_a^*/E_{\mathrm{NMR}}) \tag{3.4.9}$$

leading again to a Meyer–Neldel rule [Krühler et al., 1984] with a large scatter of the data [Schauer and Smeškal, 1987]. Note that with a sweep of V one can vary the activation energy and thus obtain the Meyer–Neldel rule for a single sample as in the case of the Staebler–Wronski effect.

An interpretation of the data of Fig. 3.17 in terms of densities of states is impeded by the following reasons:

i) The scaling law (3.4.7) is generally not fulfilled. We have already mentioned that for thin samples the conductivity of the samples is thickness dependent. For the thick samples where this effect is of minor importance the range of thicknesses that can be used to check the scaling law is unsufficient.
ii) The Meyer–Neldel rule must be understood first in order to correlate correctly the prefactor changes and the simultaneous changes of the apparent activation energy with the quasi-Fermi energy.
iii) It is always assumed that there are no problems with sample homogeneity or with contacts. We shall demonstrate below that there is evidence for a sample inhomogeneity on a mesoscopic scale which strongly influences the current voltage characteristics.

If we evaluate the SCLC in the standard way for a homogeneous sample with a density of states distribution that rises exponentially according to

$$g(E) = g(E_F) \exp\left(\frac{E - E_F}{E_T}\right) \tag{3.4.10}$$

with some characteristic energy E_T we obtain values for $g(E_F)$ in the 10^{16} to $10^{17}\,\mathrm{cm^{-3}\,eV^{-1}}$ range with $E_T \simeq 0.1\,\mathrm{eV}$.

3.5 Field Effect

One major application of amorphous a-Si:H is certainly the thin-film field effect transistor for large area integrated electronics. It is, therefore, quite natural that this device structure has attracted the interest of many research groups. Furthermore the investigation of the field effect can provide information about the microscopic transport parameters and at the same time about the density of states distribution around the equilibrium Fermi energy. For a long time the only reliable quantitative determination of the density of states in the pseudogap was by means of the field effect [Spear and LeComber, 1972].

A typical thin-film field effect transistor is depicted schematically in Fig. 3.18. The sample consists of a Metal Insulator Semiconductor (MIS) structure. The current I_{SD} from contacts called source and drain driven by the voltage V_{SD} is controlled by the gate voltage, V_G, that induces charge in a thin surface layer just beneath the insulating layer. Figure 3.19 shows typical results for the (Ohmic, i.e., linear with respect to V_{SD}) current as a function of the gate voltage (for a review see [Schumacher et al., 1988]; see also [Jang and Lee, 1983; Šmíd et al., 1985; Schropp et al., 1986, 1987; Kagawa and Muramatsu, 1986; Schumacher et al., 1987; Djamdji and LeComber, 1987]).

The interpretation of the field effect is at first sight straightforward. The induced charge condenses primarily into states above the equilibrium Fermi level. The carrier statistics is, therefore, given again by a quasi-Fermi level. In contrast to the SCLC experiment, however, we must take into consideration that the charge distribution in space is inhomogeneous, being confined to a narrow region near the insulator interface.

For a given gate voltage the current as a function of the temperature is found to be activated. Figure 3.20 shows a sequence of Arrhenius plots for the current I_{SD} for different gate voltages. Again the intercept of the current extrapolated

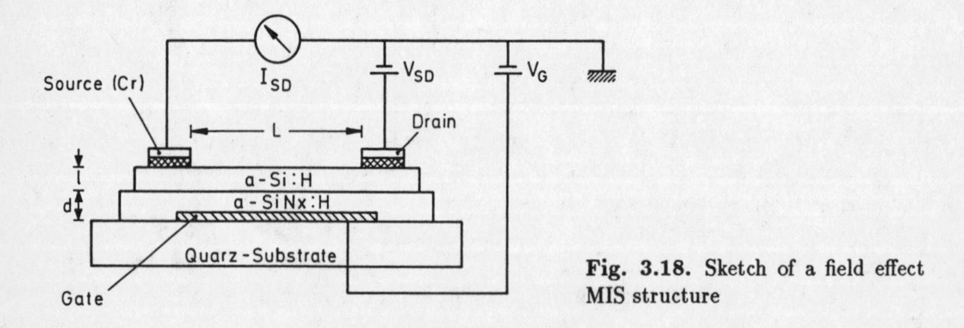

Fig. 3.18. Sketch of a field effect MIS structure

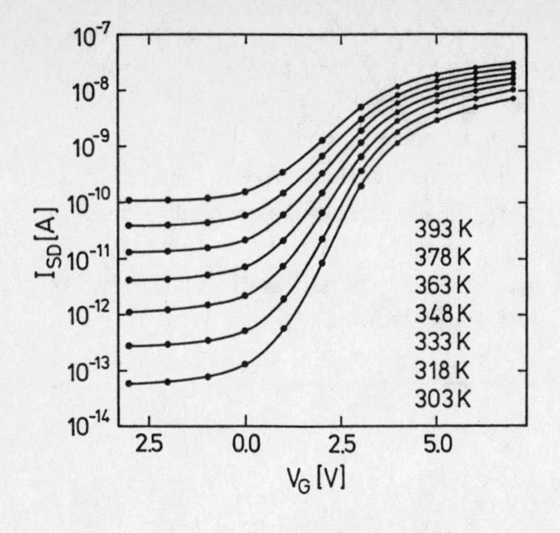

Fig. 3.19. Current versus gate voltage for an n-type a-Si : H field effect sample (from [Schumacher et al., 1987])

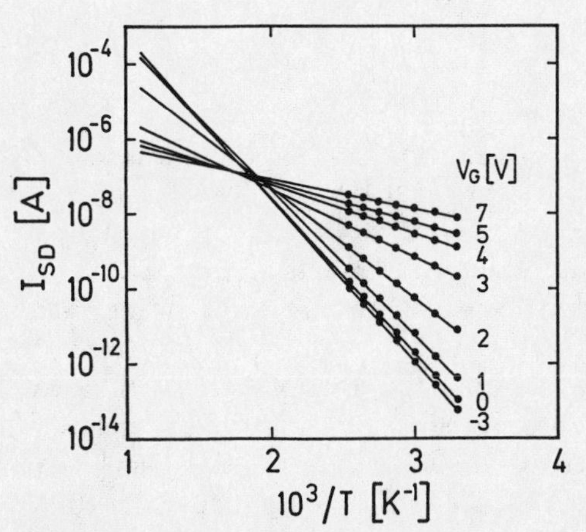

Fig. 3.20. Meyer–Neldel rule for the field effect current with the gate voltage as curve parameter

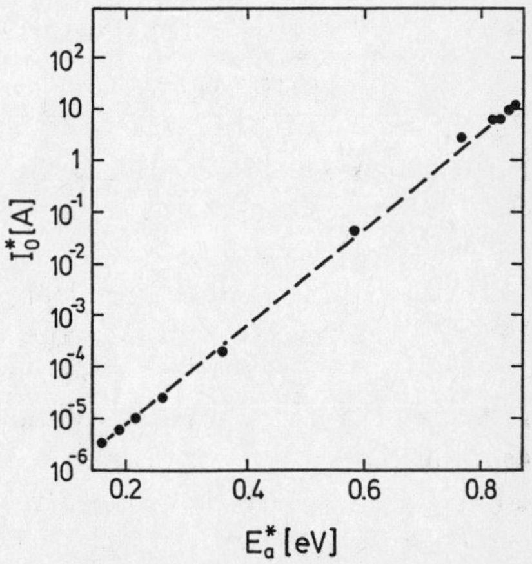

Fig. 3.21. The apparent prefactor I_0^* versus activation energy E_σ^* from the data of Fig. 3.19 (from [Schumacher et al., 1987])

to $1/T = 0$, I_0^*, for a single MIS device depends on the apparent slope, E_a^*, in the form of a Meyer–Neldel rule shown in Fig. 3.21. The interpretation of this latter effect is somewhat involved and must be postponed until Sect. 7.1.4 where we discuss the field effect in more detail.

3.6 Time-of-Flight Experiments

The time-of-flight (TOF) experiment differs fundamentally from all experiments discussed so far since it is concerned with carriers that are far from thermal equilibrium. The experimental setup is shown schematically in Fig. 3.22: The sample is in a sandwhich configuration with blocking contacts. Carriers of both polarities are created as a thin sheet below the transparent electrode using a flash of highly absorbed radiation. The external field, V_D, moves carriers of one polarity into the transparent electrode while the carriers of the opposite polarity are dragged through the sample [Spear, 1957].

While these carriers are moving through the sample, a current, I_D, is induced in the external circuit which exactly matches the internal current (provided the dielectric relaxation time is sufficiently large). We, therefore, expect a current-time relation as sketched in Fig. 3.23: After a very short initial pulse of ambipolar conduction we have a time interval with a constant current until, at $t = t_\tau$, the transit time, the carriers are collected by the electrode at the substrate.

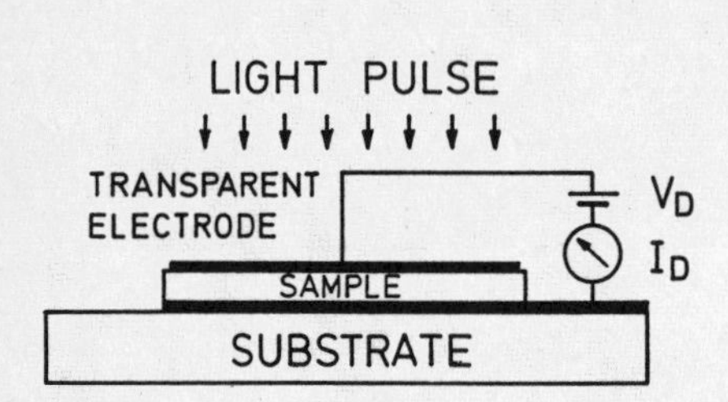

Fig. 3.22. Sketch of the setup for a time-of-flight experiment

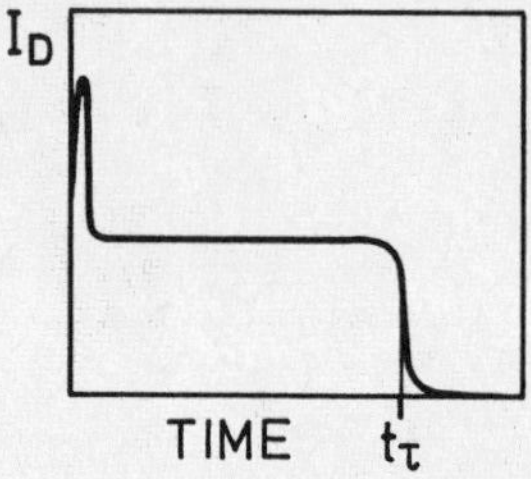

Fig. 3.23. The expected current transient for an ideal TOF experiment

If the carriers experience trapping and release events during transit the current-time relation will be more complicated. There are two limiting cases: If the average carrier gets frequently trapped and released by shallow traps the mean velocity of the carriers will be reduced but the sheet of carriers will arrive as a coherent pulse which is somewhat broadened. The transit time, therefore, is no longer a measure of the mean velocity of the free carriers in this case [Spear, 1969]. In the other extreme the carriers are trapped by deep traps which have release times long compared to the timescale of the experiment. In this case the current will decay exponentially and the transit of the few carriers that have not experienced trapping may not be observed.

Time-of-flight experiments in amorphous semiconductors [LeComber and Spear, 1970; for a review, see e.g., Marshall, 1983; Tiedje, 1984] quite generally show a $I(t)$ characteristic that is quite different from both limiting cases. The observed transients show no break that could be identified with a transit time nor an exponential decay but at first sight they appear to be rather featureless. A plot of log(current) versus log(time), however, reveals a break in the curve marking some sort of a transit time [Scharfe, 1970]. In most experiments (see Fig. 3.24 which gives a most beautiful example for the algebraic power laws measured by Tiedje [1984]) it is observed that the current decreases following a power law

$$I(t) = t^{-(1-\alpha_1)} \tag{3.6.1}$$

up to the break point, called the "transit time" after which the current is given by

$$I(t) = t^{-(1+\alpha_2)} \ . \tag{3.6.2}$$

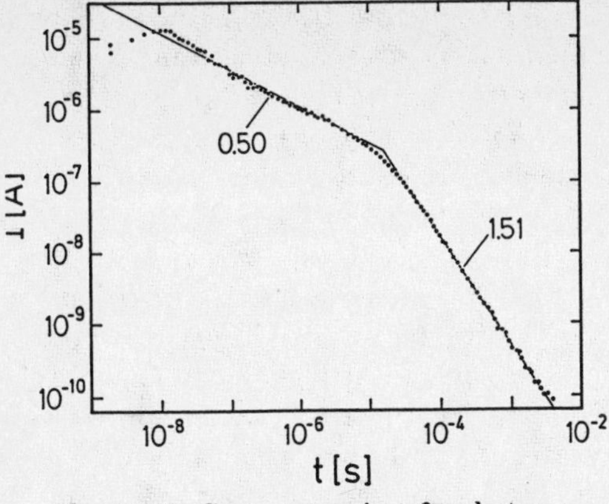

Fig. 3.24. Current transient for electrons in a-Si : H at $T = 160\,\mathrm{K}$ (from [Tiedje, 1984])

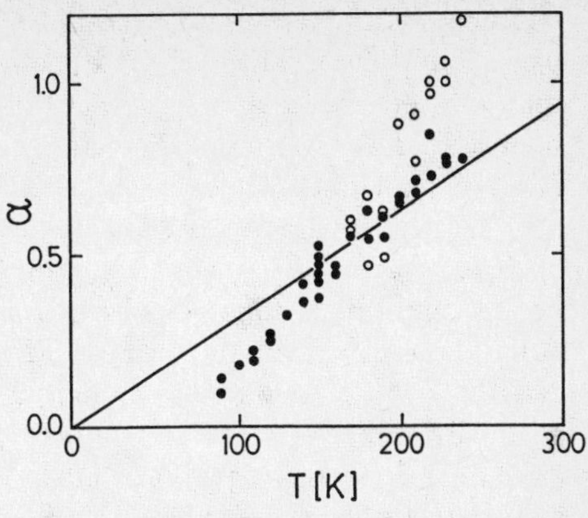

Fig. 3.25. Temperature dependence of the dispersion parameter α_1 (*solid circles*) and α_2 (*open circles*) (from [Tiedje, 1984])

Current transients of the form (3.6.1) and (3.6.2) are called dispersive or anomalously dispersive. While from the first experiments it was concluded that $\alpha_1 \cong \alpha_2$ and independent of temperature it is now established that $\alpha_1 \neq \alpha_2$ with a temperature dependence as given in Fig. 3.25. While α_1 is roughly proportional to T this relation certainly does not hold for α_2.

If one defines the break point in Fig. 3.24 as the transit time and derives a drift mobility from the transit time using (2.6.7) one finds an activated mobility with activation energies of 0.15 eV for electrons and about 0.3 eV for holes in a-Si : H. These values, however, depend on the applied field: It is a characteristic feature of the anomalous dispersive transport that there is a superlinear dependence of the drift velocity upon the applied field.

The first theoretical description of this phenomenon was given by Scher and Montroll [1975]: anomalous dispersion was concluded to be caused by a very broad distribution of characteristic delay times experienced by the carriers during transit. This is a general feature found in all models for anomalously dispersive transport. Pollak [1977] showed that R hopping would not lead to sufficient dispersion (at that time both values of α where considered temperature independent and, therefore, a temperature independent mechanism for the dispersive transport was required) while multiple trapping in an exponential tail could explain the data. This model was also proposed independently by Marshall [1977], by Schmidlin [1977], by Silver and Cohen [1977], and by Noolandi [1977]. It was further discussed in several papers and its approximate solution is now usually referred to as the Tiedje, Rose [1980], Orenstein and Kastner [1981], the "TROK" model of multiple trapping: carriers assumed to move in extended states with the mobility μ_0 are trapped in an exponential distribution of traps all with essentially the same capture cross section. While release from shallow traps is fast, carriers will be trapped and released several times by these traps. Trapping by the less abundant deep traps is rather rare but connected with release times that can exceed the time scale of the experiment.

This multiple trapping mechanism leads to a "degradation" of the carrier packet. Initially most carriers are mobile or in shallow traps. Later on more and more carriers are trapped in deeper traps and hence the average release times are increased. This is a natural explanation both for the shape of the current pulse and for the field dependence of the drift mobility derived from these pulses.

While the simplest evaluation of this TROK model leads to $\alpha_1 \cong \alpha_2 \cong T/T_{\mathrm{TOF}}$ where kT_{TOF} equals the exponential slope parameter E_{T} of (3.4.10) one can show in a Monte Carlo experiment that this needs a slight modification. After the break point the current first decays with an exponent α_2 that is not exactly equal to α_1 but at very long times the slope of the log-log plot changes gradually and the very last deeply trapped carriers give rise to a power law with an exponent equal to α_1. A forcefit to the data around the break point, therefore, does not give the exponent expected for very long times. For electrons in a-Si : H one derives from the slopes α_1 that $T_{\mathrm{TOF}} \cong 300$ K [Tiedje, 1984].

It should be noted that there is a number of alternative explanations for the observation that the exponent α_2 is not strictly proportional to T. These include nonexponential trap distributions [Silver et al., 1981; Michiel et al., 1983] and energy dependent trapping cross sections. If both the temperature dependence and the field dependence of the current transients are evaluated properly one derives a distribution of the trapping centers that decays more rapidly than an exponential tail both for the conduction band tail [Marshall et al., 1986] and for the valence band tail [Marshall et al., 1988]. The mobility of the electrons in extended states thus obtained is with $20\ \mathrm{cm^2\,V^{-1}\,s^{-1}}$ by a factor of 2 smaller than the respective value for holes.

There is of course always the possibility that the low temperature transients are due to localized carriers hopping in tails. Grünewald et al. [1985] have worked out the theory for this case which has independently been proposed by Monroe [1985].

There has been some discussion if the observed transients are due to drifting carriers or caused by an optically induced rearrangement of the space charge in a depletion layer [Silver et al., 1982, 1983]. From the observation that the current is reversed by a reversal of the applied field [Spear and Steemers, 1983] one concludes that the former interpretation is correct.

While for holes in a-Si : H there is general agreement about the anomalous dispersive transients there is a strange discrepancy between different experimental groups in the case of electrons in a-Si : H. The Dundee group reports nondispersive transients down to 150 K whereas most other groups report dispersive behaviour already below room temperature. It is not clear why such a discrepancy of experimental results from different groups persists.

In a recent paper Street et al. [1988] have shown that the high temperature value of the drift mobility can also be determined from a comparison of the dc conductivity data with the density of occupied tail states. Since this latter quantity can be measured by sweep-out experiments the drift mobility can thus be determined for the highly conducting doped samples.

3.7 ac Conductivity

A strong frequency dependence of the conductivity was reported by Pollak and Geballe [1961] for hopping conduction in impurity states of doped crystalline semiconductors. For a system that exhibits Variable Range Hopping, Austin and Mott [1969] showed that the ac conductivity in the high frequency limit could be calculated in the pair approximation

$$\sigma(\omega) = \frac{\pi}{3}\, e^2 (g(E_\mathrm{F}))^2\, \xi_{\mathrm{loc}}^5\, k\,T\,\omega(\ln(\omega_0/\omega))^4 \tag{3.7.1}$$

where ω_0 is a frequency of the order of a phonon frequency and ξ_{loc} is again the decay length of the localized states at the Fermi energy. This result was also obtained by Pollak [1971] on the basis of percolation theory. Experimentally one finds in unhydrogenated amorphous semiconductors where Variable Range Hopping prevails that

$$\sigma(\omega) \sim \omega^s \tag{3.7.2}$$

with an exponent $s \simeq 0.8$, which is practically undistinguishable from (3.7.1). For lower frequencies one finds a deviation from (3.7.2) in the dc limit. From the weak temperature dependence of the ac conductivity as compared to the dc conductivity one understands that the frequencies at which the transition from the dc to the ac limit occurs must depend on temperature also. It is clear that a pair approximation cannot cover this transition properly. The theory by Movaghar et al. [1980, 1981] describes successfully the entire frequency range.

For undoped hydrogenated a-Si:H we show in Fig. 3.26 data for the real part of the ac conductivity as a function of the temperature for two different

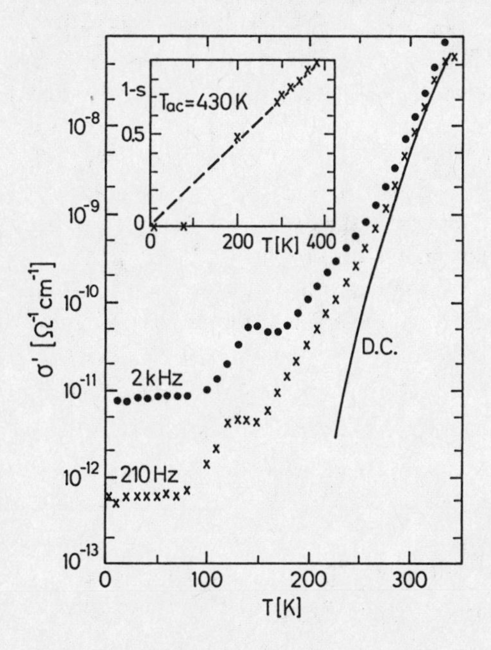

Fig. 3.26. ac conductivity for an intrinsic a-Si:H film as a function of the temperature. The insert shows the frequency exponent, $1 - s$ (from [Shimakawa et al., 1987])

frequencies [Shimakawa et al., 1987]. For the lowest temperatures there is an ac conductivity that is perfectly independent of the temperature (there is not even the linear temperature dependence of (3.7.1)) and the conductivity is proportional to the frequency. For temperatures above 300 K the conductivity is practically independent of the frequency and equals the dc conductivity. There is a remarkably wide range of temperatures where the ac conductivity is orders of magnitude above the dc conductivity and strongly temperature dependent. In this range the exponent s varies with temperature as shown in the insert

$$s = 1 - T/T_{\mathrm{ac}} \tag{3.7.3}$$

with $T_{\mathrm{ac}} \simeq 430$ K (the peak in the conductivity curves at about 150 K is due to an n^+ layer on top of the sample).

The experiments, in particular the temperature dependence of the transition from a dispersive ac current to a nondispersive dc current suggest that both transport regimes are directly linked. The frequency dependence of the ac conductivity can be converted by means of a Fourier transform into the time dependence [Butcher, 1985]

$$I(t) \sim t^{-s} \tag{3.7.4}$$

which should hold for a current caused by a voltage step and which looks quite similar to the dispersive transients discussed in the previous section. It must be noted, however, that the ac conductivity experiment differs fundamentally from the TOF experiment: while the latter experiment is performed using excess carriers that cease to contribute as soon as they thermalize the former experiment deals with carriers in thermal quasiequilibrium exclusively. In both experiments the observed dispersion requires the presence of a wide distribution of characteristic times. In the TOF experiment the standard assignment of this distribution is the distribution of detrapping times. This mechanism, however, does not lead to dispersion in the ac case. A likely interpretation of both experiments by a single mechanism in a homogeneous system would be hopping in tails [Grünewald et al., 1984, 1985]. The different values of the characteristic temperatures in both experiments, T_{ac} and T_{TOF}, could well be caused by the fact that one experiment deals with carriers in quasiequilibrium while the other one monitors excess carriers (or that the samples in both experiments are slightly different). An interpretation in terms of an inhomogeneous model will be given in Sect. 6.4.3. The temperature independent ac conductivity at very low temperatures seems to be related to deep dangling bond states, since intense illumination causes an increase of this conductivity by a factor of 2 [Shimakawa et al., 1987].

3.8 Summary of the Transport Experiments

We can summarize the experimental findings as follows: In all experiments where our Standard Transport Model would predict an activated behaviour with activation energies $E^*_{\mathrm{act}} = E_{\mathrm{C}} - E_{\mathrm{F}}$ or $E^*_{\mathrm{act}} = E_{\mathrm{C}} - E^{\mathrm{q}}_{\mathrm{F}}$, respectively, we find a

Meyer–Neldel behaviour. The Meyer–Neldel parameters for the various experiments (dc transport of a sequence of differently doped samples, dc transport for samples at different light soaked states, space charge limited current experiments with different applied voltages, field effect experiments with different gate voltages) are not exactly equal. In all cases, however, the spread of the values for the prefactor, σ_0^*, extends over several orders of magnitude and is largest for samples where the density of midgap states is expected to be rather small.

For the relatively few samples for which the temperature dependence of the thermoelectric power has been published we find for this quantity a similar rule for the activation energies and the extrapolated intercepts at $1/T = 0$. It is further striking that any deviation from a strictly activated behaviour of the conductivity is mirrored by the thermoelectric power.

The Q function which is a combination of the two aforementioned transport coefficients exhibits an entirely different variation: Again the plot versus inverse temperature shows activated behaviour. But now the intercept at $1/T = 0$, Q_0^*, is remarkably insensitive to all procedures that can alter the sample properties including a variation of the chemical composition and of the sign of the carriers and hardly ever deviates from $Q_0^* = 10 \pm 1$. The slope, E_Q^*, depends somewhat on the particular sample properties. In particular the value of E_Q^* usually is increased by any treatment that increases the disorder of the sample (light exposure, extended annealing, or electron bombardment, e.g.).

In the next two chapters a microscopic theory for transport in a homogeneous disordered system will be developed including the electron-phonon coupling. This theory is found to support and justify our Standard Transport Model of the last chapter with some modifications. In particular, the microscopic prefactor will turn out to be a nearly universal quantity and the theory does not allow for the large variations of the apparent prefactors of the conductivity observed experimentally.

4. A Transport Theory for a Homogeneous Model

In the next four chapters we shall develop a microscopic theory of transport in disordered semiconductors. Here we start by considering a microscopic model which is homogeneous on a mesoscopic scale, i.e. the disorder potential fluctuations are short-range correlated. We ignore long-range correlated random potentials. Only the conduction band is treated and transport is due to electrons. The corresponding theory for hole conduction follows by symmetry.

In Chap. 2 we have developed a Standard Transport Model for activated transport on the basis of simple physical arguments. We have in particular made use of the concept of Anderson localization, which leads to a sharp mobility edge in the absence of inelastic processes and at $T = 0$. In amorphous semiconductors, however, at $T = 0$, an electron brought into a state close to the mobility edge will lose energy by emitting phonons. Moreover, equilibrium transport in semiconductors can only be measured at sufficiently high temperatures where the absorption of phonons becomes possible as well. As a consequence inelastic scattering and incoherent hopping processes have to be considered. In this situation localization does no longer exist in a strict sense. We therefore ask to what extent the Standard Transport Model has to be revised.

We will find that our theory gives a justification for most features of the Standard Transport Model. In addition, however, some modifications will be necessary, which lead us to a slightly revised Standard Transport Model.

It has been shown in Sect. 2.8 that the transport coefficients σ and S of a semiconductor can be expressed in terms of the differential conductivity $\sigma(E)$, provided that the inelastic processes involved at the transport channel at E do not transfer energies larger than about kT. The differential conductivity $\sigma(E)$ is calculated in the next two chapters for a model which is based on the Anderson Hamiltonian (2.4.1) augmented by terms containing the phonon system and its coupling to the electron system. We shall treat the dynamic aspects of the electron-phonon interaction implying transfer of energy from the phonon system to the electron system. We will argue that the static aspects, i.e., lattice relaxation at occupied states are of minor importance.

In this chapter we outline the theory in order to show the physical contents and omit the technical details. We discuss the results for the differential conductivity, $\sigma(E)$, as a function of energy and calculate the transport coefficients $\sigma(T)$ and $S(T)$ for the semiconductor for some typical model parameters.

4.1 Construction of an Effective Single Band Model

We now specify our model and introduce the relevant microscopic parameters. In the n-type semiconductor considered here the current is carried by electrons in the lower part of the conduction band, i.e. in states close to the mobility edge. At least in the low temperature limit transport will be due to hopping. We, therefore, prefer to use the site representation where hopping processes can most naturally be described. Scattering processes as, e.g., electron-phonon scattering can of course be treated more easily in the k representation. In the situation at hand, however, we may neglect the electron-phonon scattering as a limiting process for transport in comparison to the scattering at the static disorder. The site representation is in particular useful if a tight-binding situation is considered, i.e., the coupling between the states at different sites is in some sense small. We think that as far as the lower part of the conduction band is considered (or the upper part of the valence band), this is a valid assumption.

It is generally assumed [Soukoulis et al., 1984, 1985; Economou et al., 1987] that the states in the energy range considered are due to atomic clusters which are characterized by a fluctuation of the short-range atomic arrangement. As an example we have discussed in Sect. 2.2 the fluctuation of the dihedral angle of a double tetrahedron, the fluctuation of the potential inside the interstitial holes, etc. We identify the sites of the model with the atomic clusters that give rise to the states at the band tails, e.g., with the double tetrahedra and interstitial holes. These clusters, when isolated, have eigenstates $|i\rangle$ with fluctuating eigenenergies ε_i. They will be distributed according to a distribution function $P(\varepsilon_i)$ and we take

$$\langle\varepsilon_i\rangle = 0 \tag{4.1.1}$$

$$\langle\varepsilon_i\varepsilon_j\rangle = \eta^2\,\delta_{ij} \quad . \tag{4.1.2}$$

The sites are on the average a distance a apart. For simplicity we shall place the sites on a simple cubic lattice with lattice constant a. Consequently we neglect disorder in the transfer matrix elements J between these states, i.e., we treat the case of diagonal disorder only. In addition, we take only those transfer matrix elements to be non-zero which connect nearest neighbour states. The electronic Hamiltonian then reads

$$H_\mathrm{e} = \sum_i (\varepsilon_i - E)\,n_i + J\sum_{ij} c_i^\dagger c_j \tag{4.1.3}$$

with

$$n_i = c_i^\dagger c_i \tag{4.1.4}$$

which is the Anderson Hamiltonian (2.4.1), however, applied to a system where the sites represent atomic clusters instead of single atoms.

Let us now discuss the magnitude of the microscopic parameters. The coupling J is certainly much smaller than that of a tight-binding model describing the total width of the conduction band, as e.g., the Hall–Weaire–Thorpe model [Hall, 1952, 1958; Weaire and Thorpe, 1971] where it is of the order of some eV. In our model J should be smaller, we assume it to be typically 0.1 eV which leads to a tight-binding band with a total width of $12\,J \approx 1$ eV. If we take the disorder parameter η to be close to J, there will be two mobility edges, E_{C} and E'_{C}, separating localized states at the band edges from delocalized states at the central part. We will be interested in the transport properties of this band if the energy E is in its lower part.

It remains to specify the lattice constant a. This will certainly be larger than the interatomic distance. A value of $a = 5\,\text{Å}$ seems to be reasonable. The density of the basis states $|i\rangle$ is then $8 \times 10^{21}\ \mathrm{cm}^{-3}$.

So far we have discussed the electronic part of the model. In addition we shall consider phonons and the coupling of electrons to phonons. The phonon Hamiltonian will be written

$$H_{\mathrm{p}} = \sum_{\boldsymbol{q}} \hbar\omega_{\boldsymbol{q}} \left(b_{\boldsymbol{q}}^{\dagger} b_{\boldsymbol{q}} + \tfrac{1}{2} \right) \tag{4.1.5}$$

where we still use the classification of phonon modes in terms of $\boldsymbol{q}$, see Sect. 2.3.

In the approach used here the phonon system enters only in the form of frequency convolutions over phonon and electron propagators. It is not essential how the phonon states are described in detail since only their spectral distribution matters. For this function we shall use a model which is technically convenient.

The electron-phonon coupling will be introduced by modelling the coupling of the phonon system to the basis states $|i\rangle$ at the sites of our electronic Hamiltonian H_{e}. We consider only the diagonal coupling to the density n_i and neglect nondiagonal terms $\sim c_i^{\dagger} c_j$, which are exponentially smaller. This is a general practice in hopping theories and has been introduced already by Miller and Abrahams [1960]. The electron-phonon coupling constant $A_{i\boldsymbol{q}}$ is a free parameter of the model. Here we shall not treat small polaron effects, and will therefore assume $A_{i\boldsymbol{q}}$ to be sufficiently small. For the interaction term we write

$$H_{\mathrm{ep}} = \sum_{\boldsymbol{q}} (A_{i\boldsymbol{q}}\, b_{\boldsymbol{q}}^{\dagger} + A_{i-\boldsymbol{q}}\, b_{\boldsymbol{q}})\, n_i \quad . \tag{4.1.6}$$

Electron-electron interaction need not be taken into account since we are describing a semiconducting system with a very low density of carriers in the band.

Our model is now complete. It is expressed by the model Hamiltonian

$$H = H_{\mathrm{e}} + H_{\mathrm{ep}} + H_{\mathrm{p}} \quad . \tag{4.1.7}$$

In the following section we shall derive a formula for the frequency dependent conductivity within standard linear response theory on the basis of our model. This expression will be the starting point of our theory. In Sect. 4.3 the mode-coupling approach to Anderson localization will be presented for the phonon-free case. The applicability of these results to amorphous semiconductors will be questioned in detail in Sect. 4.4. We then proceed by incorporating the electron-phonon interaction in the limiting case of strongly localized states which leads to hopping conduction. It is then shown how the mode-coupling approach to Anderson localization can be generalized in order to describe also the effects due to the electron-phonon interaction for electron states close to the mobility edge.

4.2 The Conductivity

Our approach is based on a determination of the dc limit of $\sigma(z)$, the complex conductivity for the model presented above. $\sigma(z)$ is a function of the complex frequency z and of the energy E, which will be identified with the Fermi energy of a *degenerate* electron system. Since electron-phonon interaction is now included $\sigma(z)$ is a generalization of (2.8.11b). Once this function is determined, we obtain the transport coefficients of the semiconductor using the Kubo–Greenwood formulae (2.8.12, 14). As discussed in Sect. 2.8 this procedure is a reasonable approximation. In the dc-limit we will write $\sigma(z \to 0) = \sigma(E)$ which will be referred to as the *differential conductivity.*

The conductivity is the response to an external electric field $\boldsymbol{F}(t)$ introducing a perturbation, which in the sense of our tight-binding model is taken to be

$$H' = -\sum_j V_j n_j \quad . \tag{4.2.1}$$

The external potential

$$V_j(t) = e\boldsymbol{R}_j \cdot \boldsymbol{F}(t) \tag{4.2.2}$$

couples to the carrier density, leading to a density fluctuation described by the expectation value $\langle \delta n_i(t) \rangle$. The expectation value of the current $\langle \boldsymbol{j}(t) \rangle$ is then given by the time derivative of the polarization, or in frequency space

$$\langle \boldsymbol{j}(z) \rangle = -\mathrm{i}z \frac{e}{\Omega} \sum_i \boldsymbol{R}_i \langle \delta n_i(z) \rangle \tag{4.2.3}$$

where Ω is the volume of the system and $\boldsymbol{R}_i$ the position of site i.

From standard linear response theory we have

$$\langle \delta n_i(z) \rangle = \sum_j \chi_{ij}(z) V_j(z) \tag{4.2.4}$$

with the susceptibility (taking $\hbar = 1$)

$$\chi_{ij}(t) = \tfrac{1}{2} \langle [n_i , n_j(-t)] \rangle \tag{4.2.5}$$

which has the Laplace transform

$$\chi_{ij}(z) = 2\mathrm{i} \int_0^\infty \mathrm{e}^{\mathrm{i}zt} \chi_{ij}(t)\, dt \quad . \tag{4.2.6}$$

The indices i, j refer to a pair of sites, while the time evolution is determined by the whole quantum mechanically coupled system of sites.

An amorphous system is isotropic on the average. We may therefore replace the tensor $\boldsymbol{R}_i\boldsymbol{R}_j$, which appears in (4.2.3) when (4.2.2) and (4.2.4) are inserted, by $\boldsymbol{R}_i\cdot\boldsymbol{R}_j/3$ times the unit tensor. Because of particle conservation $[H, \sum_i n_i] = 0$ and hence $\sum_j \chi_{ij}(z) = 0$, we can therefore replace $\boldsymbol{R}_i\cdot\boldsymbol{R}_j/3$ by $-R_{ij}^2/6$, where $R_{ij}^2 = (\boldsymbol{R}_i - \boldsymbol{R}_j)^2$.

The conductivity then follows as

$$\sigma(z) = -\frac{e^2}{6\Omega}(-\mathrm{i}z)\sum_{ij} R_{ij}^2\,\chi_{ij}(z) \tag{4.2.7}$$

which is a complex function of the complex frequency z. We will postpone the precise formulation of the theory until the next chapter, where $\sigma(z)$ will be calculated using (4.2.7) on the basis of our model. Here we only mention that $\chi_{ij}(z)$ can be related to the quantum mechanical current-current correlation function, leading to the Kubo formula for the conductivity. Alternatively $\chi_{ij}(z)$ can be written in terms of the density-density correlation function. This expresses the generalized Einstein relation between conductivity and diffusivity.

4.3 Anderson Localization

At $T = 0$ and if we neglect the coupling to phonons, our system shows Anderson localization. In this particular case our transport theory will yield two mobility edges E_C and E'_C, such that $\sigma(E) = 0$ for $E < E_\mathrm{C}$, $E > E'_\mathrm{C}$, while $\sigma(E)$ is nonzero for $E_\mathrm{C} < E < E'_\mathrm{C}$, c.f. Fig. 2.2.

The approach to Anderson localization presented here has been introduced by Götze [1978, 1979, 1981, 1985] and Belitz and Götze [1983]. We express $\sigma(z)$ in terms of a current-current correlation function $\psi(t)$, which describes current fluctuations. The current fluctuations decay with a current relaxation rate τ_j^{-1}. If this rate does not depend on frequency, the current-current correlation function $\psi(t)$ decays exponentially like $\exp(-t/\tau_j)$. In general the current relaxation is described by a kernel depending on frequency: $\tau_j^{-1}(z)$. In fact, there exists an exact expression relating any correlation function to a frequency dependent relaxation kernel. The conductivity is given by the Laplace transform of the current-current correlation function, which in general can be written as

$$\psi(z) = \frac{\mathrm{i}}{z + \mathrm{i}\tau_j^{-1}(z)}\,\psi(t=0) \quad , \tag{4.3.1}$$

hence

$$\sigma(z) = \frac{ne^2}{m}\,\frac{\mathrm{i}}{z + \mathrm{i}\tau_j^{-1}(z)} \quad . \tag{4.3.2}$$

For a frequency independent rate τ_j^{-1} this reduces to the Drude formula, i.e., for the real part of $\sigma(z \to \omega^+)$ we get

$$\sigma'(\omega) = \frac{ne^2}{m} \frac{\tau_j^{-1}}{\omega^2 + \tau_j^{-2}} \tag{4.3.3}$$

where n and m are the density and the mass of the carrier, respectively.

Since the conductivity is related to the diffusivity, we can alternatively consider the density-density correlation function $\phi(t)$. This quantity can also be written in the frequency domain as

$$\phi(z) = \frac{\mathrm{i}}{z + \mathrm{i}\tau_n^{-1}(z)} \phi(t=0) \tag{4.3.4}$$

which again is an exact expression.

If the density relaxation rate does not depend on frequency the density fluctuations again decay exponentially like $\exp(-t/\tau_n)$. The density relaxation rate τ_n^{-1} is then related to the diffusion coefficient D. In general $\tau_n^{-1}(z)$ is given by the correlation function of the forces acting on the density fluctuations, i.e., of the currents. We therefore have

$$\tau_n^{-1}(z) \sim \psi(z) = \frac{\mathrm{i}}{z + \mathrm{i}\tau_j^{-1}(z)} \psi(t=0) \quad , \tag{4.3.5}$$

which is a generalized Einstein relation between conductivity (ψ) and diffusivity (τ_n^{-1}).

It was Götze's observation that $\tau_j^{-1}(z)$ can be approximately expressed by $\phi(z)$

$$\tau_j^{-1}(z) = \nu_0 + \alpha\, \phi(z) \tag{4.3.6}$$

where α is a constant containing the model parameters and the density of states at the Fermi energy, while ν_0 is a zeroth order current relaxation rate which can be identified with the frequency independent τ_j^{-1} of (4.3.3) and determines the zeroth order approximation σ_{00} to the conductivity

$$\sigma_{00} = \frac{ne^2}{m} \frac{\mathrm{i}}{z + \mathrm{i}\nu_0} \quad . \tag{4.3.7}$$

This closed set of equations (4.3.4, 5, 6) exhibits interesting properties. It describes feedback in the following sense. Consider the limit $z \to 0$. If the carrier density fluctuations at the energy E slow down ($\tau_n^{-1}(z \to 0)$ is small), the elastic scattering at the static disorder becomes more effective ((4.3.4, 6): $\phi(z \to 0)$ and $\tau_j^{-1}(z \to 0)$ become large). As a consequence, the fluctuations move even slower (4.3.5), etc. The strength of this feedback mechanism depends on energy E. By lowering the energy, starting from the center of the band, eventually at a certain energy $E = E_C$ scattering becomes so effective that the density fluctuations are no longer able to decay, they become stationary: E_C is the mobility edge. The same happens mutatis mutandis when approaching the upper band edge.

More formally this behaviour is given by the analytic properties of the function $\tau_j^{-1}(z)$. If it is an analytic function for $z \to 0$, then the density fluctuations decay and the dc conductivity is non-zero. On the other hand, if $\tau_j^{-1}(z)$ exhibits a pole like i/z, then the current-current correlation function vanishes in the dc limit and the conductivity is zero. At the same time the density-density correlation function diverges as i/z, i.e., is constant in time. There is no diffusion in the dc limit, $\tau_n^{-1}(z \to 0) \to 0$, i.e., no density relaxation.

It will be shown in the next chapter that the resulting dc conductivity can be written as [see also Belitz et al., 1981]

$$\sigma(E) = \sigma_{00} \cdot (1 - y) \tag{4.3.8}$$

where y is a coupling constant containing the constant α, and σ_{00} is the zeroth order conductivity defined in (4.3.7). Interestingly enough it turns out that the product $\sigma_{00} y = \sigma_M = 0.03 \times e^2/\hbar a$, which is a value close to Mott's [1972] minimum metallic conductivity (2.4.7). We will later argue that this value for σ_M is nearly universal.

Figure 4.1 shows results of a model calculation, where the density of states distribution of the band has been taken to be equal to the distribution function $P(\varepsilon_i)$ of site energies, which was box-shaped in this particular case: $P(\varepsilon_i) = 1/W$ for $|\varepsilon_i| \leq W/2$. The dotted line is σ_{00} as a function of energy E. The horizontal broken line represents the value σ_M, while the solid line is the conductivity $\sigma(E) = \sigma_{00} - \sigma_M$, showing two mobility edges. One can read off that close to E_C the following expression holds to a good approximation (if energies are measured from the lower band edge)

$$\sigma(E) = \sigma^{(0)} \cdot \left(\frac{E - E_C}{E_C}\right)^{\lambda} \tag{4.3.9}$$

with the critical exponent $\lambda = 1$ and $\sigma^{(0)} \approx \sigma_M$. These results substantiate the predictions of the scaling theory discussed in Sect. 2.4.2. They are also the basis of our Standard Transport Model introduced in Sect. 2.6.

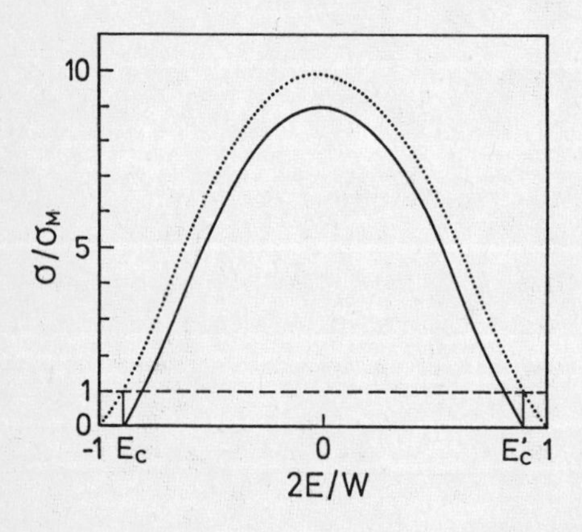

Fig. 4.1. $\sigma(E) = \sigma_{00} - \sigma_M$ for a box-shaped band. The dotted line is σ_{00}, the dashed line is σ_M

4.4 Problems with the Localization Picture for Amorphous Semiconductors

The Anderson localization picture has been widely applied for the interpretation of experimental transport data in the amorphous semiconductor field. If there are two mobility edges, one in the conduction band and one in the valence band, the well-established approaches used for crystalline semiconductors seem to be applicable. There are several reasons, however, that this picture is an oversimplification and therefore unsatisfactory.

It is known, that at any non-zero temperature transport between localized states can take place via hopping processes. These hopping processes can be described by transition rates which increase with increasing localization length. Scaling theory tells us that the localization length diverges for energies approaching the mobility edge. In addition the mobility edges are presumably located in energy ranges with a high density of states. The spatial density of the relevant localized states is then high and their mutual distance is small. Consequently, the contribution to the current from carriers moving by hopping close to the mobility edges may well become comparable to the contribution from carriers in delocalized states. These considerations show that the mobility edge loses its meaning as a boundary between conducting and non-conducting states at non-zero temperatures. Even at $T = 0$, in a semiconductor, energy relaxation of an electron injected into the band occurs via hopping, i.e., we have transport below E_C.

More fundamentally, the whole concept of localization becomes meaningless at finite temperatures if there are inelastic processes [Thouless, 1977; Imry, 1980]. To get a feeling of what happens we sketch contour plots of wave functions just below and just above the mobility edge, Figs. 4.2, 3. The eigenfunctions have fractal character in both energy regimes close to E_C [Aoki, 1982, 1983; Soukoulis and Economou, 1984]. On a length scale which is smaller than ξ_{loc} they look rather similar to each other. Inelastic processes scatter carriers from one state to another at a certain rate. The dwelling time of a particle in an eigenstate may then be so small that the particle is not able to realize whether the state is in fact localized or delocalized. In other words, the finite lifetime in a particular eigenstate makes the distinction between localized and delocalized states meaningless.

In the language of scaling theory (c.f. Sect. 2.4.2) the conductivity in a system without interactions at $T = 0$ is characterized by the correlation length ξ which diverges at the mobility edge, c.f. (2.4.28). If we introduce interactions, then at E_C the relevant length is the ineleastic scattering length L_i. This length is determined by the diffusion of the particle which takes place on a microscopic length scale and is related to the microscopic conductance g_3, where the β function, (2.4.11), vanishes. At the mobility edge the conductivity on macroscopic length scale L is then given by (2.4.29), where ξ has been replaced by L_i. Since this latter length is finite the conductivity at the mobility edge is finite, too. For

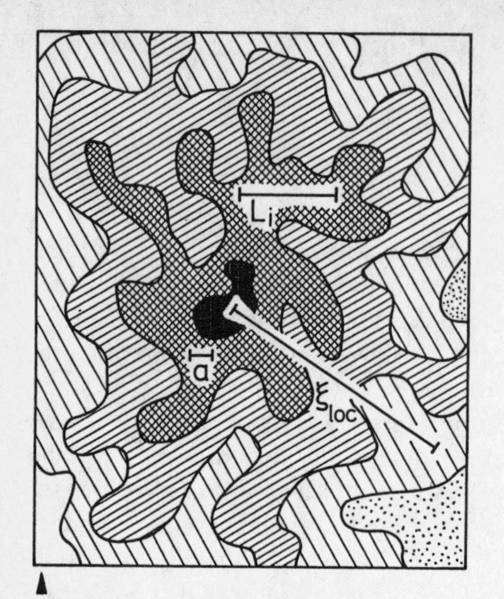

Fig. 4.2. Artists impression of a contour plot of $|\psi(r)|^2$ for an energy E slightly below E_C such that $\xi_{loc} \gg a$. On a length scale exceeding the nearest neighbour site distance a the wave function ψ has a non-uniform shape. An inelastic scattering length $a < L_i < \xi_{loc}$ is also indicated

Fig. 4.3. The same plot as in Fig. 4.2, however for an energy E slightly above E_C

energies below the mobility edge in the noninteracting system the conductivity is zero and the relevant length is the localization length ξ_{loc}. If for the interacting system the inelastic scattering length L_i is smaller than ξ_{loc} (see Fig. 4.3), the distinction between localized and delocalized states becomes meaningless. There will be no qualitative difference in the propagation of an electron above or below E_C. This effect will be referred to as *phonon induced delocalization*, a notion which to our knowledge has first been used by Kikuchi [1983].

Before we discuss in more detail how the electron-phonon interaction modifies the conductivity also outside the critical region, we consider in more detail the case of hopping transport, which has been introduced already in Chap. 2 on a phenomenological basis. This will enable us to show that our transport theory reduces to conventional hopping theory in the appropriate limit.

4.5 The Rate Equation Formulation of Hopping Theory

In this section we introduce the electron-phonon coupling and consider first transport in an energy range where we can be sure that the electronic eigenstates are sufficiently localized. As an example consider impurity bands, the dangling bond band or the lower part of the band tail of amorphous semiconductors. We describe hopping in terms of the transfer rates W_{ij} which have been introduced in Sect. 2.7. These are given by the square of the matrix element of H_{ep} taken between the two localized electronic eigenstates considered in the elementary hopping process [Miller and Abrahams, 1960]. They also contain energy conservation. We will refer to these rates as *binary rates.* If there exists an external electric field, the rates will depend also on the potential V_j. The same is true for the expectation value of the occupation number

$$f_i(t) = \langle n_i(t) \rangle \quad . \tag{4.5.1}$$

According to (4.2.3) we need $\delta f_i(z)$ which can be determined from a phenomenological rate equation describing gain and loss of the occupation at site i

$$\frac{\partial}{\partial t}\, \delta f_i(t) = \sum_j \{ f_j (1 - f_i)\, W_{ji} - f_i (1 - f_j)\, W_{ij} \} \quad , \tag{4.5.2}$$

with

$$\delta f_i(t) = f_i(t) - f_i^0 \quad . \tag{4.5.3}$$

Here

$$f_i^0 = f_{\mathrm{F}}(\varepsilon_i, E_{\mathrm{F}}, T) \tag{4.5.4}$$

is the equilibrium Fermi distribution at energy ε_i. The rate equation can be linearized [Brenig et al., 1971, 1973], yielding in frequency space

$$-\mathrm{i}\, z\, \delta f(z) = -CF^{-1}\, \delta f(z) - \beta C V(z) \tag{4.5.5}$$

where a matrix notation is understood. F is a vector with components F_i, given by

$$F_i = f_i^0 (1 - f_i^0) \tag{4.5.6}$$

and C is the collision matrix

$$C_{ij} = -\Gamma_{ij}^0 + \sum_l \Gamma_{il}^0\, \delta_{ij} \quad . \tag{4.5.7}$$

The symmetric equilibrium hoprates Γ_{ij}^0 have been introduced already in (2.7.3). They include electron statistics

$$\Gamma_{ij}^0 = W_{ij}^0\, f_i^0 (1 - f_j^0) = \Gamma_{ji}^0 \quad . \tag{4.5.8}$$

W_{ij}^0 is the transfer rate without external field (2.7.2).

The linearized rate equation (4.5.5) has the formal solution (with $\beta = 1/kT$ for the rest of this book)

$$\delta f(z) = -\mathrm{i}\, \beta (z + \mathrm{i}\, CF^{-1})^{-1}\, CV(z) \tag{4.5.9}$$

which, if inserted into (4.3.2) for $\langle \boldsymbol{j}(z) \rangle$ gives the hopping conductivity. We obtain

$$\sigma(z) = -\left\langle \frac{e^2 \beta}{6\Omega} \sum_{ij} R_{ij}^2 \left\{ \frac{z}{z + \mathrm{i}\, CF^{-1}} C \right\}_{ij} \right\rangle_{\mathrm{config}} \quad . \tag{4.5.10}$$

The major task, however, has still to be worked out, namely the configurational averaging with respect to the distribution of the ε_i (and of the distances, if the sites are randomly distributed), denoted by $\langle\ \rangle_{\text{config}}$. This can be done using the mean field approach developed by Movaghar and coworkers [Movaghar et al., 1980a, 1980b, 1981; Movaghar and Schirmacher, 1981]. The results agree well with those of resistor network calculations and percolation theory applied to the rate equation [McInnes et al., 1980].

Equation (4.5.10), when written as an infinite series, expresses the hopping conductivity in terms of all possible sequences of binary hopping events connecting sites i and j. These hopping events are described by binary rates, containing only information about the two sites and their coupling. This is considered to be a valid approximation if the hopping events are rather rare. Close to a mobility edge, however, the localization length increases and the density of states is large. Consequently the hopping rates increase drastically. We then expect that the inelastic scattering length L_{i} becomes smaller than ξ_{loc}, and that transport should better be described on the basis of quantum mechanical tunneling. This mode of propagation, however, is precluded by the binary rate equation approach since in this limit phase coherence is destroyed after each single hop. Propagation via tunneling, on the other hand, is related to a strict phase memory of the wave function from site to site. In order to retain also these latter processes in a more complete theory one has to go beyond the binary limit. Only in the appropriate limit of strongly localized states the theory should approach the hopping expression (4.5.10).

4.6 Phonon Induced Delocalization

Let us now try to bridge the gap between the localization theory and the binary hopping theory. We turn back to Sect. 4.3 and consider what happens if the electron-phonon coupling is switched on. This interaction induces density relaxation, i.e. transport, for all positions of the Fermi energy E in the degenerate system, in particular for E below the mobility edge E_{C}, as has been discussed in the previous section. For E sufficiently above E_{C}, on the other hand, the electron-phonon interaction will reduce the conductivity due to scattering. As already mentioned, we do not take into account this latter aspect of the interaction, since for a semiconductor the contribution from the high energy range to the total conductivity will be strongly reduced due to the Boltzmann factor in the Kubo–Greenwood formula.

We shall see in the next chapter that the electron-phonon interaction can be viewed as a (dynamical) disorder of the nondiagonal elements of our tight-binding Hamiltonian. Belitz and Götze [1982] have found, by introducing *static* non-diagonal disorder in addition to the static diagonal disorder, that the former has a delocalizing effect if the diagonal disorder is strong enough. (For weak diagonal disorder, however, non-diagonal disorder leads to localization.) It is, therefore, plausible that the dynamical disorder due to the phonons will have a similar delocalizing effect provided the static disorder is large.

Since the electron-phonon interaction is dynamic, we will get an energy transfer from the phonon system to the carrier system. In the localized regime this is related to the hopping processes discussed above. We, therefore, obtain another contribution to the conductivity, and consequently to density relaxation. We shall refer to the density relaxation rate $\tau_n^{-1}(z)$ discussed in Sect. 4.4 as the (quantum mechanical) *tunneling rate* as opposed to the phonon assisted *hopping rate*. In addition we will get a hopping density relaxation rate, denoted by $h_n(z)$. This function will turn out to be a generalization of the binary collision matrix C_{ij}, (4.5.6). Approximately $h_n(z)$ is given by a frequency convolution over a phonon propagator $\Lambda(\omega)$ and the phonon-free tunneling rate $\tau_n^{-1}(z)$. It reduces to C_{ij} in the binary hopping limit. The dc limit $h_n(z \to 0)$ turns out to be non-zero for all energies in the band, i.e., there is always hopping.

The hopping density relaxation rate $h_n(z)$ appears as an additional contribution to the denominator of $\phi(z)$ in (4.3.4), i.e.

$$\phi(z) = \frac{\mathrm{i}}{z + \mathrm{i}\,\tau_n^{-1}(z) + \mathrm{i}\,h_n(z)}\,\phi(t=0) \quad . \tag{4.6.1}$$

Consequently this function never diverges in the dc limit. The same is then true for the current relaxation rate $\tau_j^{-1}(z)$, because of (4.3.6). Since in turn $\tau_j^{-1}(z)$ determines the tunneling rate $\tau_n^{-1}(z)$ (c.f. (4.3.5)), this latter function never vanishes in the dc limit as well.

The conductivity, given by $\tau_n^{-1}(z) + h_n(z)$, is a sum of a tunneling contribution and an incoherent hopping contribution. The tunneling contribution never vanishes in the dc limit in contrast to the phonon-free case, i.e. there is a tunneling contribution in addition to the hopping processes even at energies below E_C. This describes phonon-induced delocalization in the present approach. We identify the phonon-induced delocalization mechanism as being due to a breaking of the feedback mechanism, resulting from the presence of hopping-like density relaxation at all energies.

4.7 The Differential Conductivity for the Coupled Electron–Phonon System

The approach outlined above can be worked out in detail for the model introduced in Sect. 4.1 [Müller and Thomas, 1983, 1984]. As an input the following microscopic parameters are required which have been discussed earlier in this chapter:

a	lattice constant of the underlaying lattice
J	transfer matrix element
A_{iq}	electron-phonon coupling strength
η^2	the disorder parameter for the site energies.

In addition we have to introduce three model functions:

$\Lambda(\omega)$ the phonon propagator
$\varrho(E)$ the density of electronic states per site
$P(\varepsilon_i)$ the distribution function of the site energies.

$\Lambda(\omega)$ and $\varrho(E)$ could in principle be calculated from the model. Since $\Lambda(\omega)$ will enter under frequency integrals only its particular form does not matter much. We assume that it is given by

$$\Lambda(\omega) = \frac{\theta}{\pi} \frac{\omega_0}{\omega^2 + \omega_0^2} \tag{4.7.1}$$

where

$$\theta = \int \Lambda(\omega)\, d\omega \tag{4.7.2}$$

and ω_0 is a typical phonon frequency, the last input parameter in our theory.

This function is assumed to model the interaction with acustical phonons having a Debye frequency of the order of ω_0. Its technically convenient form has been used previously by Jonson and Girvin [1979] (see also [Girvin and Jonson, 1980]). The temperature dependence of θ is essentially determined by the Bose–Einstein distribution. We shall consider temperatures that are high enough to allow for a linear approximation: $\theta = r\,kT$. The parameter θ will be taken as a dimensionless temperature. Since we treat the electronic system using $T = 0$ statistics, temperature enters only via the combination θ. The parameter r contains the coupling A_{iq} and the phonon frequency ω_0. In the limit of single-phonon processes we require that

$$\theta < 1 \quad . \tag{4.7.3}$$

$\theta = 0$ means that the interaction is switched off.

The density of states per site, $\varrho(E)$, is determined by the magnitude of the transfer matrix element J and the distribution of site energies, $P(\varepsilon_i)$. If the latter is broad enough, i.e. η comparable to J as assumed here, $\varrho(E)$ can be approximated by $P(\varepsilon_i)$, or vice versa. We will choose various model functions for $P(\varepsilon)$ describing the lower half of the tight-binding band with its tailing. From those functions η is calculated.

The ratio η/J is chosen such that a mobility edge at $\theta = 0$ appears at a reasonable position close to the band tail. The lattice constant a enters only in the final result for the conductivity. The phonon frequency ω_0 will be fixed to be 0.05 eV throughout. Its particular value turns out to have only little influence on the final results.

The only really free parameter is the electron-phonon coupling strength represented by r. We have no crystalline reference to determine how large the electron-phonon coupling constant A_{iq} should be because our sites are not the atoms of the random network. In the present model we only assume that $\theta < 1$, which for the temperature range 200 K $< T <$ 600 K sets an upper limit for r.

The differential conductivity $\sigma(E)$ has been calculated [Fenz et al., 1985] using various models for the density of states $\varrho(E)$

$$\varrho(E) = \varrho_0 \begin{cases} \exp[-(-E/E_\mathrm{T})^n] & E < 0 \\ 1 & E_\mathrm{max} > E > 0 \end{cases} \quad n \geq 1 \tag{4.7.4}$$

where E_max is a cutoff energy and E_T the decay parameter of the tail. The limit $n \to \infty$ corresponds to a box-shaped band, i.e., no tails at all. In that case we take $E_\mathrm{max} = W$, the width of the distribution. Remember that the conductivity for a semiconductor is obtained by a convolution of $\sigma(E)$ with a Boltzmann distribution. The contributions from energies larger than $E_\mathrm{max} \simeq kT$ will be negligible. The prefactor ϱ_0 is determined by the normalization

$$1 = \int \varrho(E)\, dE \quad . \tag{4.7.5}$$

The following general behaviour of $\sigma(E)$ is observed on a logarithmic scale (see Fig. 4.4 for an illustration): For E approaching E_C from above $\sigma(E)$ first decreases slightly. It has a value of the order of σ_M there. For $\theta = 0$ it vanishes at E_C, which corresponds to a sharp bending down on the logarithmic scale. If θ is not too small, the function $\sigma(E)$ tends to follow the density of states in the high density tail region. Only for small θ there appears a kink, showing a distinct transition from purely incoherent hopping to tunneling behaviour. This kink is smeared out at elevated temperatures, showing the effect of the phonon induced delocalization.

This latter point becomes clearer if we look at Fig. 4.5, where the total differential conductivity and its hopping contribution are shown for the box-shaped

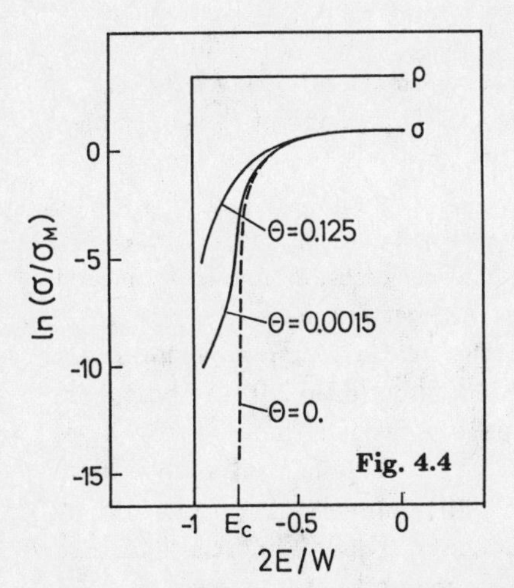

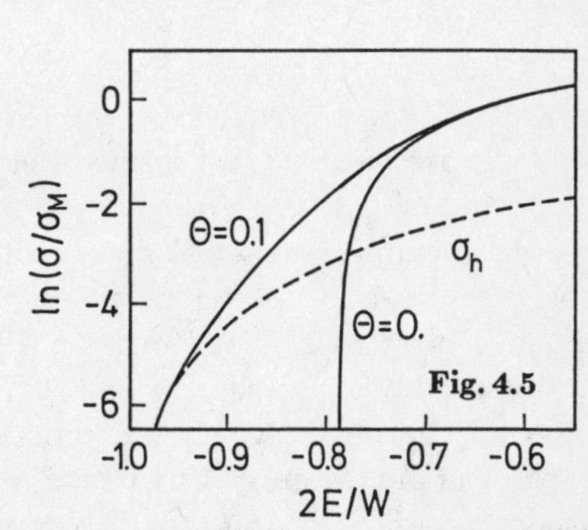

Fig. 4.4. Logarithmic plot of the differential conductivity $\sigma(E)$ for the model underlying Fig. 4.1 and for three different values of the parameter θ. For $\theta = 0$ we obtain the mobility edge E_C. For nonzero values of θ the differential conductivity does not vanish below E_C

Fig. 4.5. Enlarged plot of the differential conductivity $\sigma(E)$ as in Fig. 4.4. For $\theta = 0.1$ the hopping contribution is indicated by the dashed line

band (see also Fig. 4.1 for the case $\theta = 0$). We see that just below E_C the total conductivity is much higher than the pure hopping contribution. This indicates, that for the parameters chosen ($\theta = 0.1$) a two path model is not applicable.

4.8 Transport Coefficients for the Semiconductor Model

The transport coefficients for the semiconductor, i.e. σ, S, and their combination Q, are obtained by inserting $\sigma(E)$ into the Kubo–Greenwood formulae, (2.8.12, 14). However, before doing this, we may draw some qualitative conclusions. The integrand of (2.1.14) has a maximum at a certain energy, called the dominant transport energy E_t here, which is implicitly given by

$$\frac{d \ln \sigma(E_t)}{dE} = \beta = \frac{1}{kT} \quad . \tag{4.8.1}$$

For $\theta = 0$, i.e. in the localization case, E_t is slightly above E_C and does not vary much with temperature, because the curvature of $\sigma(E)$ is rather sharp there. If we consider $\theta > 0$, and not too small, this bending over becomes more gradual due to the phonon induced delocalization. Consequently the transport energy E_t will in general have a more pronounced temperature shift. In addition, the function $\sigma(E)$ itself depends on temperature, leading to an increase of $\sigma(E)$ close to E_C and an increased tailing with rising temperature. Let us define a temperature coefficient of the transport energy by (taking only the first order of the Taylor expansion)

$$E_t(T) = E_t^*(0) + \gamma_t^* T \quad . \tag{4.8.2}$$

Here $E_t^*(0)$ is the extrapolated value of $E_t(T)$ for $T \to 0$. The shift term γ_t^* may in principle be positive or negative, depending on the interplay between the phonon induced increase of $\sigma(E)$ and the decrease of the slope of the Boltzmann factor.

For very low temperatures $\sigma(E)$ developes a kink separating pure hopping processes at low energies from tunneling processes at higher temperatures. We will then have two maxima in the integrand of (2.8.14), in particular if the density of states (and hence $\sigma(E)$) level off at lower energies, as is the case in an amorphous semiconductor. One maximum is then close to E_F and dominates transport at the lowest temperatures, while the contribution from the second maximum close to E_C is strongly reduced due to the Boltzmann factor. If we assume in particular that the tail of $\sigma(E)$ is purely exponential with a slope parameter E_T below E_C, the dominant transport channel will jump from the lower maximum to energies close to E_C, if kT exceeds E_T. If T increases further, a weak shift of $E_t(T)$ to lower or higher energies may occur.

In our model calculations we have found that usually γ_t^* is negative for sufficiently high temperatures. Its value has been found not to exceed k typically and to depend to some degree both on the shape of the density of states close to E_C and on the electron-phonon coupling parameter r.

The conductivity $\sigma(T)$ of the semiconductor calculated from $\sigma(E)$ using the Kubo–Greenwood formula (2.8.14) has been found to follow an activated behaviour [Fenz et al., 1985]

$$\sigma(T) = \sigma_0 \exp(-\beta E_\sigma^*) \quad . \tag{4.8.3}$$

The resulting thermoelectric power can be approximated by

$$(q/k)\, S(T) = A^* + \beta E_S^* \tag{4.8.4}$$

with a well defined activation energy E_S^*. The Q function then reads

$$Q(T) = \ln(\sigma_0\, \Omega\, \mathrm{cm}) + A^* - \beta E_Q^* \tag{4.8.5}$$

with

$$E_Q^* = E_\sigma^* - E_S^* \quad . \tag{4.8.6}$$

According to our convention to assign all quantities with an asterisk which are derived from Arrhenius plots we should have written σ_0^* instead of σ_0. It will be shown in Chaps. 6, 7 that the experimental prefactor σ_0^* differs from that found here due to processes not yet included in our model. We, therefore, denote the prefactor resulting from the present homogeneous model by the symbol σ_0. We call σ_0 the *microscopic prefactor* of the conductivity since it is determined by processes taking place on a microscopic length scale.

It turns out that the value of σ_0 is close to σ_M. Values for particular models will be given below. For a first estimate we may for a moment ignore the phonon coupling and take the expression (4.3.9) for $\sigma(E)$. Then we find that $\sigma_0 = \sigma_M/\beta E_C$. Since E_C is some kT above the lower band edge, we get a certain fraction of σ_M. This value is then enhanced by the shift term $\exp(-\gamma_t/k)$, $-1 < \gamma_t/k < 0$. For this particular choice for $\sigma(E)$ we obtain $E_\sigma^* = E_S^*$.

One could also try to characterize σ_0 by the value of $\sigma(E)$ at the dominant transport energy E_t. Since this energy depends on temperature according to (4.8.2), we could write $\sigma_0 = \sigma(E_t) \exp(-\gamma_t/k)$ with $-1 < \gamma_t/k < 0$. This procedure would make sense only if $\sigma(E)$ were to rise sharply for energies approaching the mobility edge from below. But if we look at Figs. 4.4 and 4.5 we see that near E_C the differential conductivity $\sigma(E)$ has a smooth variation there leading to a rather broad and in general asymmetric maximum of the integrand of the Kubo–Greenwood formula. In other words, there is a broad range of energies contributing to the conductivity of the amorphous semiconductor.

Let us now turn to the results of model calculations [Fenz et al., 1985] based on the density of states models introduced in the last subsection. We find that in general the Q function has a finite activation energy when plotted against T^{-1}, typically $0.05\,\mathrm{eV} < E_Q^* < 0.1\,\mathrm{eV}$. Only for the phonon-free case we have essentially horizontal Q functions. The best defined E_Q^*, i.e. straight lines for Q vs. T^{-1}, are obtained for steep tails. In particular for the box shaped band we find (taking $W = 1\,\mathrm{eV}$, $J = 0.125\,\mathrm{eV}$, $r = 3.2\,\mathrm{eV}^{-1}$)

$$\sigma_0 = \sigma_M \tag{4.8.7a}$$

$$A^* = 2.2 \tag{4.8.7b}$$

$$E_Q^* = 0.06\,\text{eV} \tag{4.8.7c}$$

$$Q_0^* = 2.2 + \ln(\sigma_M\,\Omega\,\text{cm}) = 7.2 \quad . \tag{4.8.7d}$$

This value for Q_0^* results if we take the lattice constant to be 5 Å, i.e. $\sigma_M \simeq 150\Omega^{-1}\,\text{cm}^{-1}$. Similar values have been obtained for less steep tails. It was found, however, that the tailing parameter should not exceed a value of $E_T = 170\,\text{K}$ (for a purely exponential tail), if we want to describe straight Q functions.

For a non-exponential and smoother tail ($n = 4$, c.f. (4.7.4)) we obtain (taking $E_T = 0.2\,\text{eV}$, $J = 0.06\,\text{eV}$, $r = 4.5\,\text{eV}^{-1}$)

$$\sigma_0 = 2.2\,\sigma_M \tag{4.8.8a}$$

$$A^* = 2.5 \tag{4.8.8b}$$

$$E_Q^* = 0.07\,\text{eV} \tag{4.8.8c}$$

$$Q_0^* = 3.3 + \ln(\sigma_M\,\Omega\,\text{cm}) = 8.3 \quad . \tag{4.8.8d}$$

The important point here is that the value of σ_0 is rather independent of details of the model. Remember that $\sigma_M = 0.03\,e^2/\hbar a$, where the number 0.03 is only weakly model dependent (see the next chapter) and a is some Å. In particular, σ_M does not depend on the density of states distribution close to E_C. This function rather determines the ratio σ_0/σ_M, which varies by a factor of typically 2 from model to model. We, therefore, avoid to use the concept of a mobility when describing the microscopic transport processes close to a mobility edge in an equilibrium situation. We rather propose that the microscopic prefactor could be regarded as a nearly universal quantity describing transport in all such cases where the static disorder leads to a dominant transport path close to a mobility edge. The density of states distribution function mainly determines the position of the mobility edge in the band, but has little influence on the value of the differential conductivity $\sigma(E)$ there.

If we compare E_σ^* with $E_C - E_F$ we find that

$$\Delta E_\sigma^* = E_\sigma^* - (E_C - E_F) \tag{4.8.9}$$

varies between $-50\,\text{meV}$ and $+50\,\text{meV}$, depending on the model of the density of states used. For the box shaped band we find $\Delta E_\sigma^* = -0.05\,\text{eV}$, while for the case $n = 4$ we get $\Delta E_\sigma^* = +0.05\,\text{eV}$.

The theory sketched so far has made use of a number of non-trivial approximations (see Chap. 5). It is, therefore, desirable to test the theory by comparing the results to those of a different approach. This has been done by calculating numerically the differential conductivity on a Bethe–lattice using the same Hamiltonian (4.1.7) [Dersch and Thomas, 1985] as in our present theory. The results have been found to support the analytic theory in detail. Here we omit the presentation of this approach since it does not give additional insight into the physical processes involved.

4.9 Enhanced Localization Due to Static Lattice Relaxation

As has been mentioned in Sect. 4.4 the wave functions just above E_C have a fractal character. Assume that this regime extends up to an energy $E_0 > E_C$. On the basis of the scaling theory for small polaron formation developed by Emin and Holstein [1976] it has been argued by Cohen, Economou and Soukoulis [1983, 1984] that these states have an enhanced instability against small polaron formation at $T = 0$. These highly inhomogeneous states fill only a small fraction of space and, therefore, offer little resistance against lattice relaxation if occupied. They become strongly compressed and form small polarons.

For energies higher up in the band above E_0, on the other hand, the states lose their fractal character and become stable against small polaron formation. The authors then suggest that at $T = 0$ electrons in states between E_C and E_0 become localized small polarons, i.e. the mobility edge is shifted upwards to E_0.

An extreme view has been taken by Viščor [1983] following Anderson's [1972] suggestion. He assumes that the small polaron binding energy is so large that the states at the band edge are in fact transposed right to the middle of the pseudogap.

It has not been examined to what extent the small polaron formation process survives if in addition the dynamical aspects of the electron-phonon coupling are taken into account at relatively high temperatures. At present we argue that due to phonon induced processes the fractal nature of the states close to E_C is smeared out and that small polaron formation becomes less important. In other words, the lifetime of a particle in a state close to E_C is too small to induce lattice relaxation. In this situation it is the inelastic length L_i rather than the correlation length ξ which has to be considered in the theory of Cohen et al. From the value of $\sigma(E)$ for $E = E_C$ we estimate L_i to be rather small, of the order of several lattice constants a. At this length scale the fractal character should not yet manifest itself.

Nevertheless some effect of lattice relaxation could survive. To clarify this point one has to await a complete theory of electron-phonon interaction close to a mobility edge, taking dynamic as well as static aspects of the interaction into account simultaneously.

4.10 The Revised Standard Transport Model

In summarizing this chapter we have obtained the following model for the charge transport in a homogeneous amorphous semiconductor. The conductivity σ and the thermopower S are given by (4.8.3) and (4.8.4), respectively. The microscopic prefactor σ_0 is a nearly universal quantity. Its value should be close to $150\ \Omega^{-1}\,\mathrm{cm}^{-1}$. There is an uncertainty by a factor of the order of unity partly due to the uncertainty of the lattice constant, i.e., the average distance between sites, and partly due to a residual influence of the shape of the density of states distribution which enters the temperature dependence of $\sigma(E)$ below E_C. The activation energies of conductivity and thermopower, respectively, differ by at least $E_Q^* = 0.05$ eV. A value exceeding about 0.1 eV, however, can not be explained on the basis of the homogeneous model treated so far. The finite value of E_Q is due to the phonon induced delocalization which produces a tailing of the differential conductivity $\sigma(E)$ below E_C. The activation energy of the conductivity is given by $E_C - E_F$ with an uncertainty of ± 50 meV.

We thus have essentially reproduced the features of the Standard Transport Model of Sect. 2.6 with the additional important modification that 50 meV $< E_\sigma^* - E_S^* <$ 100 meV. These results will hold for systems with linear dimensions exceeding the inelastic scattering length L_i. From $\sigma(E_C)$ this length can be estimated to be some lattice constants a (i.e. less than 100 Å at the temperatures envisaged). The temperature dependence of L_i turns out to be $\sim T^{-0.6}$ [Dersch, 1986]. In contrast to the ubiquitious Meyer–Neldel rule the microscopic theory does not allow for a spread of the σ_0 values by several orders of magnitude, in fact not even for a spread that exceeds a factor of 2.

5. Detailed Presentation of the Theory

In this chapter we present in detail the theory outlined above. The interaction-free case is treated first and Götze's mode-coupling theory of localization is presented in a representation which formally resembles the hopping theory. Although we shall concentrate on the conducting phase here we also derive some results for the insulating phase. The mean square displacement is introduced, which leads to a physically appealing definition of the localization length. These considerations allow to introduce the basic idea of the potential well analogy approach. We then proceed to study the modifications due to the inclusion of the electron-phonon coupling.

Our main task will be the development of a theory for the differential conductivity $\sigma(E)$. Remember that the Kubo–Greenwood formula (2.8.11a, 11b, or 14) expresses the temperature dependent conductivity $\sigma(T)$ of the semiconductor in terms of a function $\sigma(E)$ which is the conductivity, at $T = 0$, of the system if the Fermi energy is situated within the band at the energy E. If there are no inelastic processes this description is exact. Here, however, we shall also consider the influence of inelastic processes on $\sigma(E)$. The treatment is then approximative (see Sect. 2.8). We shall take the electron subsystem at $T = 0$, while the phonon subsystem is considered at finite temperatures. We shall consider three-dimensional systems exclusively.

Throughout this chapter we shall take $\hbar = 1$. We start with a tutorial survey of correlation functions, where we follow the notation of Forster's textbook [1975].

5.1 Correlation Functions

Consider an external perturbation

$$H' = -B\,V(t) \tag{5.1.1}$$

where $V(t)$ is an external potential coupling to an observable B. The expectation value of an observable A, $\langle A(t)\rangle$, is then given within linear response theory by

$$\langle A(t)\rangle = \int_{-\infty}^{t} \chi_{AB}(t-t')\,V(t')\,dt' \quad . \tag{5.1.2}$$

Let us assume without loss of generality that $\langle A\rangle = \langle B\rangle = 0$. The susceptibility χ (we omit the indices A, B) is given by

$$\chi(t) = \langle [A^{\dagger}, B(-t)]\rangle/2 \quad . \tag{5.1.3}$$

It has a Fourier transform, the spectrum

$$\chi''(\omega) = \int_{-\infty}^{\infty} \mathrm{e}^{\mathrm{i}\omega t}\, \chi(t)\, dt \quad , \tag{5.1.4}$$

which is related via the Dissipation–Fluctuation Theorem

$$\chi''(\omega) = (1 - \mathrm{e}^{-\beta\omega})\, S(\omega)/2 \tag{5.1.5}$$

to the Van Hove correlation function

$$S(\omega) = \int_{-\infty}^{\infty} \mathrm{e}^{\mathrm{i}\omega t}\, S(t)\, dt \quad , \tag{5.1.6}$$

with

$$S(t) = \langle A^{\dagger}\, B(-t) \rangle \quad . \tag{5.1.7}$$

The $\langle \cdots \rangle$ denotes thermodynamic averaging over the equilibrium system and may or may not include the configurational average over the static disorder.

Consider a non-ergodic system, where $S(t) \to$ const., if $t \to \infty$. An example is a density fluctuation which will not decay in time if we consider the Fermi level E to be in the localized regime. This produces a $\delta(\omega)$ contribution in the spectrum $S(\omega)$. According to (5.1.5) this $\delta(\omega)$ contribution is missing in $\chi''(\omega)$. Therefore we prefer to express $\sigma(z)$ in terms of correlation functions.

It is convenient to include the quantum factor, $1-\exp(-\beta\omega)$, appearing in the Fluctuation–Dissipation Theorem, into the definition of a quantum mechanical correlation function, the so-called Kubo correlation function. It is defined by

$$\frac{\mathrm{i}\beta}{2}\, \partial_t\, \phi(t) = \chi(t) \quad , \tag{5.1.8}$$

which reads for the Fourier transforms

$$\phi''(\omega) = \int_{-\infty}^{\infty} \mathrm{e}^{\mathrm{i}\omega t}\, \phi(t)\, dt = \frac{2}{\beta\omega}\, \chi''(\omega) \quad . \tag{5.1.9}$$

There may be, however, an additional $\delta(\omega)$ contribution on the RHS of this equation. Then the system is called non-ergodic.

We define the Laplace transforms of these functions by ($\mathrm{Im}\, z > 0$)

$$\chi(z) = 2\mathrm{i} \int_{0}^{\infty} \mathrm{e}^{\mathrm{i}zt}\, \chi(t)\, dt \quad , \tag{5.1.10}$$

$$\phi(z) = \int_{0}^{\infty} \mathrm{e}^{\mathrm{i}zt}\, \phi(t)\, dt \quad . \tag{5.1.11}$$

The Hilbert–Stieltjes transform relates the Laplace transform to the spectra

$$\chi(z) = \int_{-\infty}^{\infty} \frac{d\omega}{\pi}\, \frac{\chi''(\omega)}{\omega - z} \tag{5.1.12}$$

$$\phi(z) = \frac{1}{2i} \int_{-\infty}^{\infty} \frac{d\omega}{\pi} \frac{\phi''(\omega)}{\omega - z} \tag{5.1.13}$$

with the typical causal pole which is the Fourier transform of the step function $\Theta(-t)$.

From (5.1.8) and (5.1.10) we have

$$\chi(z) = -\beta \int_0^{\infty} e^{izt}\, \partial_t \phi(t)\, dt \tag{5.1.14}$$

$$= \beta\phi(t=0) + iz\beta\, \phi(z) \quad . \tag{5.1.15}$$

Thus

$$\phi(z) = \frac{\chi(z) - \beta\, \phi(t=0)}{iz\beta} \tag{5.1.16}$$

has a singularity like i/z for $z \to 0$ if the dc limit of $\chi(z)$ does not agree with $\beta\, \phi(t=0)$. This singularity corresponds to the $\delta(\omega)$ contribution in $\phi''(\omega)$, according to (5.1.13).

An explicit expression for $\phi(t)$ is:

$$\phi(t) = \frac{1}{\beta} \int_0^{\beta} \langle A^\dagger\, B(-t + i\lambda)\rangle\, d\lambda \quad . \tag{5.1.17}$$

One can easily show that its Fourier transform is

$$\phi''(\omega) = \frac{1 - e^{-\beta\omega}}{\beta\omega}\, S(\omega) \quad , \tag{5.1.18}$$

which gives

$$\phi''(\omega) = \frac{2}{\beta\omega}\, \chi''(\omega) \tag{5.1.19}$$

because of the Fluctuation–Dissipation Theorem. In the classical limit $\beta\omega \to 0$ we get

$$\phi''(\omega) = S(\omega) \quad . \tag{5.1.20}$$

Equation (5.1.17) can be interpreted as a scalar product in the so-called Liouville space:

$$\phi(t) = (A \,|\, B(-t)) \tag{5.1.21}$$

$$= (A \,|\, e^{-iLt} \,|\, B) \tag{5.1.22}$$

where

$$\dot{B} = iLB = i\,[H, B] \quad . \tag{5.1.23}$$

We also have (c.f. (5.1.11))

$$\phi(z) = \left(A \left| \frac{\mathrm{i}}{z - L} \right| B\right) \quad , \tag{5.1.24}$$

which has all the formal properties of a scalar product.

A useful relation follows from (5.1.3, 8, 21, 22)

$$\mathrm{i}\partial_t (A \,|\, B(-t))\Big|_{t=0} = (A \,|\, L \,|\, B) \tag{5.1.25}$$

$$= \frac{2}{\beta}\, \chi(t=0) \tag{5.1.26}$$

$$= \frac{1}{\beta}\, \langle [A^\dagger \,,\, B] \rangle \quad . \tag{5.1.27}$$

This implies that

$$(A \,|\, L \,|\, B) = 0 \tag{5.1.28}$$

whenever

$$[A^\dagger, B] = 0 \quad . \tag{5.1.29}$$

5.2 Application to the Tight–Binding Model. Phonon–Free Case

The representation of correlation functions in terms of relaxation kernels can be constructed using the Mori–Zwanzig formalism [Mori, 1965; Zwanzig, 1960, 1961]. The relaxation kernels are generalizations of the relaxation rates discussed in the previous chapter. In this section we shall treat only the electronic Hamiltonian H_{e}.

Let us first specify the operators A and B of the last section. Since $V_j(t)$ is the external potential

$$V_j(t) = e\,\boldsymbol{R}_j \cdot \boldsymbol{F}(t) \quad . \tag{5.2.1}$$

A and B are vectors with components n_j. All susceptibilities, correlation functions and relaxation kernels are therefore matrices. If no indices appear in the following a matrix notation is to be understood.

We will denote the density-density correlation function by

$$\phi_{ij}(z) = \left(n_i \left| \frac{\mathrm{i}}{z - L} \right| n_j\right) \quad . \tag{5.2.2}$$

We will also consider a "current-current" correlation function defined by

$$\psi_{ij}(z) = \left(\dot{n}_i \left| \frac{\mathrm{i}}{z - L} \right| \dot{n}_j\right) \tag{5.2.3}$$

which is related to ϕ by

$$\psi(z) = z^2 \phi(z) - \mathrm{i} z \, \phi(t=0) \tag{5.2.4}$$

as can be seen by expanding $1/(z-L)$ and using $\dot{n} = \mathrm{i}Ln$.

5.2.1 Equal-Time Correlation Functions

In (5.2.4) there appears an equal-time correlation function

$$\phi_{ij}(t=0) = (n_i \,|\, n_j) \quad . \tag{5.2.5}$$

We will also be dealing with the equal-time correlation function of $\dot{n}$

$$\psi_{ij}(t=0) = (\dot{n}_i \,|\, \dot{n}_j) \quad . \tag{5.2.6}$$

In this section we calculate these matrices approximately for the model at hand. We consider the phonon-free case and take $T=0$ at the end.

The Kubo identity

$$\frac{\partial}{\partial V} \, \mathrm{e}^{-\beta(H-BV)} \Big|_{V=0} = \mathrm{e}^{-\beta H} \int_0^\beta \mathrm{e}^{\lambda H} \, B \, \mathrm{e}^{-\lambda H} \, d\lambda \tag{5.2.7}$$

can be used to relate an equal-time correlation function $(A|B)$ to the isothermal thermodynamic susceptibility

$$\chi_T = \frac{\partial \langle A \rangle_V}{\partial V} \Big|_{V=0\,,T=\mathrm{const}} \tag{5.2.8}$$

which describes the change of an observable A with a change of the external potential V coupling to the observable B. Since

$$\langle A \rangle_V = \frac{\mathrm{Tr}\,\{\mathrm{e}^{-\beta(H-BV)} A\}}{\mathrm{Tr}\,\{\mathrm{e}^{-\beta(H-BV)}\}} \tag{5.2.9}$$

we get

$$\chi_T = \beta \, (A|B) \tag{5.2.10}$$

using (5.2.7).

Let us now consider in particular $\phi_{ij}(t=0)$. From the Hamiltonian H_e we identify V of (5.2.9) with a change δE of the Fermi energy E, coupling to $\sum_j n_j$ and have

$$\sum_j \phi_{ij}(t=0) = \sum_j (n_i|n_j) = \frac{1}{\beta} \, \frac{\partial \langle n_i \rangle_{\delta E}}{\partial \, \delta E} \Big|_{T=0\,,\delta E=0} \quad . \tag{5.2.11}$$

As an approximation we consider only the diagonal elements of $\phi_{ij}(t=0)$, i.e. the lowest order in J, and have

$$\phi_{ij}(t=0) = \frac{1}{\beta}\,\varrho\,\delta_{ij} \quad , \tag{5.2.12}$$

where ϱ is the change of particle density per site with the change of the chemical potential, i.e. it is the density of states per site at the Fermi energy.

For $\psi_{ij}(t=0)$ an explicit expression can be written down using (5.1.27)

$$\psi_{ij}(t=0) = (\dot{n}_i|\dot{n}_j) = \frac{\mathrm{i}}{\beta}\,\langle[\dot{n}_i, n_j]\rangle \tag{5.2.13}$$

with

$$\dot{n}_i = \mathrm{i}J\sum_j c_j^\dagger c_i - \mathrm{i}J\sum_j c_i^\dagger c_j \quad . \tag{5.2.14}$$

Also here we shall keep only the lowest order contribution in J, which is the nearest neighbour element of $\psi_{ij}(t=0)$ for $i \neq j$. Because of particle conservation

$$\sum_i \dot{n}_i = 0 \quad , \tag{5.2.15}$$

and therefore the diagonal matrix elements are given by

$$\psi_{ii}(t=0) = -\sum_{j(\neq i)} \psi_{ij}(t=0) \quad . \tag{5.2.16}$$

5.2.2 Representation by Single Particle Spectral Functions

For the pure electronic Hamiltonian H_e one can express a correlation function

$$\phi_{1234}(z) = \left(c_1^\dagger c_2 \left| \frac{\mathrm{i}}{z-L} \right| c_3^\dagger c_4\right) \tag{5.2.17}$$

by single particle spectral functions

$$A_{ij}(\omega) = -\mathrm{Im}\left\langle i \left| \frac{1}{\omega^+ - H_\mathrm{e}} \right| j \right\rangle \tag{5.2.18}$$

and the equilibrium Fermi distribution

$$f^0(\omega) = \frac{1}{1+\mathrm{e}^{\beta(\omega-E)}} \tag{5.2.19}$$

as

$$\phi_{1234}(z) = \frac{\mathrm{i}}{\beta}\int_{-\infty}^{\infty}\frac{d\omega}{\pi}\int_{-\infty}^{\infty}\frac{d\omega'}{\pi}\,\frac{1}{z-\omega+\omega'}\,\frac{f^0(\omega)-f^0(\omega')}{\omega'-\omega} \times \langle A_{13}(\omega)\,A_{42}(\omega')\rangle_\mathrm{conf} \tag{5.2.20}$$

using standard Green's function techniques [Kadanoff and Baym, 1962]. The configurational average $\langle\ \rangle_\mathrm{conf}$ over the product of the spectral functions appears of course only if the correlation function has been defined so as to include it. Otherwise the configurational average has to be performed at the end of the calculation.

In particular we have for the equal time correlation function

$$\phi_{1234}(t=0) = \frac{1}{2\pi} \int_{-\infty}^{\infty} \phi''_{1234}(\omega)\, d\omega \tag{5.2.21}$$

$$= \frac{1}{\beta} \int_{-\infty}^{\infty} \frac{d\omega}{\pi} \int_{-\infty}^{\infty} \frac{d\omega'}{\pi} \frac{f^0(\omega) - f^0(\omega')}{\omega' - \omega} \langle A_{13}(\omega)\, A_{42}(\omega') \rangle_{\text{conf}} \quad . \tag{5.2.22}$$

The configurational averaged density of states per site $\varrho(\omega)$ is related to the spectral function by

$$\varrho(\omega) = \frac{1}{\pi} \langle A_{ii}(\omega) \rangle_{\text{conf}} \tag{5.2.23}$$

and (c.f. (5.2.12))

$$\varrho(0) = \varrho \quad . \tag{5.2.24}$$

5.2.3 The Effective Ordered Lattice Hamiltonian

We denote the matrix formed by the non-diagonal elements of $\psi_{ij}(t=0)$ by $\tilde{h}_{ij}/\beta$. The matrix elements of $\tilde{h}$ are all zero except for the nearest neighbour elements which are all equal to a certain value denoted by $-\gamma$. The number γ can be determined approximately using (5.2.22) and decoupling the configurational average over the spectral functions. To the lowest order in J only the diagonal elements of the spectral functions contribute and we have

$$\gamma = 2J^2 \int_{-\infty}^{\infty} d\omega \int_{-\infty}^{\infty} d\omega'\, \varrho(\omega)\, \varrho(\omega') \frac{f^0(\omega) - f^0(\omega')}{\omega' - \omega} > 0 \quad . \tag{5.2.25}$$

So $\beta\psi_{ij}(t=0)$ is given by an effective tight-binding Hamiltonian $\tilde{h}$ for an ordered three-dimensional simple cubic lattice. This effective Hamiltonian has a spectrum of eigenvalues in the range $[-6\gamma\,, 6\gamma]$. The corresponding effective Green's function

$$\tilde{G}_{ij}(E) = \left(\frac{1}{\tilde{h} - E^+}\right)_{ij} \tag{5.2.26}$$

is therefore real for $E \leq -6\gamma$ and approaches its value at $E = -6\gamma$ continuously. (This is true for a three-dimensional system. In lower dimensions $\tilde{G}$ is no longer continuous at the band edge.)

Since

$$\tilde{G}_{ij}(E = -6\gamma) = \beta^{-1}\, \psi_{ij}^{-1}(t=0) \quad , \tag{5.2.27}$$

the inverse of $\psi_{ij}(t=0)$ exists in this sense. We will write

$$\tilde{h}_{ij} = \gamma\, h_{ij} \tag{5.2.28}$$

and

$$G_{ij}(E/\gamma) = \gamma \tilde{G}_{ij}(E) \tag{5.2.29}$$

and thus

$$\psi_{ij}^{-1}(t=0) = \frac{\beta}{\gamma}\, G_{ij}(-6) \tag{5.2.30}$$

with

$$G = \frac{1}{h - E^{+}} \quad . \tag{5.2.31}$$

The concept of an effective ordered Hamiltonian has been introduced previously by Movaghar and coworkers [1980, 1981] and Movaghar and Schirmacher [1981] in their formulation of the hopping theory, and by Müller and Thomas [1984] in their treatment of Anderson localization (see also a recent application by Loring and Mukamel [1986]). It also forms the basis of the potential well analogy approach introduced by Economou and Soukoulis [1983]. It is an advantage of this concept that all well known formal properties of single particle Green's functions [see Economou, 1983] can be used in a theory which actually works with two-particle Green's functions, i.e., correlation functions.

5.2.4 The Continued Fraction Representation of the Correlation Functions

A continued fraction representation of ϕ is generated by defining first projectors

$$P_0 = \sum_{ij} |n_i)\, \phi_{ij}^{-1}(t=0)\, (n_j| \tag{5.2.32}$$

and

$$Q_0 = 1 - P_0 \quad . \tag{5.2.33}$$

We then observe that the following identity holds

$$\frac{\mathrm{i}}{z - L} = \frac{\mathrm{i}}{z - Q_0 L - P_0 L} \tag{5.2.34}$$

$$= \frac{\mathrm{i}}{z - Q_0 L} + \frac{\mathrm{i}}{z - Q_0 L - P_0 L}\, P_0 L \,\frac{1}{z - Q_0 L} \quad . \tag{5.2.35}$$

The first term, when taken between $(n_i| \quad |n_j)$ gives $(\mathrm{i}/z)\, \phi_{ij}(t=0)$. All higher order terms vanish since $(n_i\,|\,Q_0 = 0$. The second term will contain $\phi(z)$ as a factor, thus we get finally

$$\phi(z) = \frac{\mathrm{i}}{z + \mathrm{i}\, \Sigma_0(z)\, \phi^{-1}(t=0)}\, \phi(t=0) \tag{5.2.36}$$

where

$$\Sigma_0(z) = \left(\dot{n}_i \,\middle|\, \frac{\mathrm{i}}{z - Q_0 L} \,\middle|\, \dot{n}_j\right) \quad . \tag{5.2.37}$$

We have used (5.1.27) and the fact that $[n_i, n_j] = 0$. $\Sigma_0(z)\,\phi^{-1}(t=0)$ is the generalized density relaxation rate. It is related to the current-current correlation function $\psi(z)$ by

$$\psi(z) = \frac{z}{z + \mathrm{i}\,\Sigma_0(z)\,\phi^{-1}(t=0)}\,\Sigma_0(z) \tag{5.2.38}$$

due to (5.2.4).

We can continue this procedure and define projectors

$$P_1 = \sum_{ij} |\dot{n}_i)\,\psi_{ij}^{-1}(t=0)\,(\dot{n}_j| \tag{5.2.39}$$

and

$$Q_1 = 1 - P_1 \quad . \tag{5.2.40}$$

This leads to a representation of $\Sigma_0(z)$ in terms of a current relaxation kernel:

$$\Sigma_0(z) = \frac{\mathrm{i}}{z + \mathrm{i}\,\Sigma_1(z)\,\psi^{-1}(t=0)}\,\psi(t=0) \tag{5.2.41}$$

with

$$\Sigma_1(z) = \left(\ddot{n}_i \,\middle|\, \frac{\mathrm{i}}{z - Q_0 Q_1 L} \,\middle|\, \ddot{n}_j\right) \quad . \tag{5.2.42}$$

Here we have used

$$(\dot{n}_i \,|\, L \,|\, \dot{n}_j) = 0 \quad , \tag{5.2.43}$$

which holds for a homogeneous equilibrium system, where $\langle\ \rangle_{\mathrm{conf}}$ is included in the definition of the correlation functions.

The matrix $\Sigma_1(z)\,\psi^{-1}(t=0)$ is the generalization of the current relaxation or scattering rate appearing in the Drude formula for the conductivity.

Instead of continuing this procedure we will truncate the continued fraction at the level of Σ_1 using a suitable approximation which leads to a selfconsistent set of equations.

5.2.5 An Approximation for the Scattering Rate

The generalized current relaxation rate is a correlation function of the forces acting on the current fluctuations. In our representation it is a correlation function of the second time derivative of the density. Writing

$$L = L_0 + L_J \tag{5.2.44}$$

where L_0 is given by the first term of H_e and L_J corresponds to the second one we have

$$\begin{aligned}\Sigma_{1ij}(z) =&\left(L_0\dot{n}_i\,\Big|\,\frac{\mathrm{i}}{z-Q_1Q_0L}\,\Big|\,L_0\dot{n}_j\right)\\&+\left(L_J\dot{n}_i\,\Big|\,\frac{\mathrm{i}}{z-Q_1Q_0L}\,\Big|\,L_J\dot{n}_j\right)\\&+\text{two cross terms}\quad .\end{aligned} \tag{5.2.45}$$

The first term of (5.2.45), denoted by Σ_1^0, describes scattering at the static disorder. In

$$L_0\dot{n}_j = \mathrm{i}J\sum_k \omega_{kj}\,(c_k^\dagger c_j + c_j^\dagger c_k) \tag{5.2.46}$$

there appears an energy difference

$$\omega_{kj} = \varepsilon_k - \varepsilon_j \tag{5.2.47}$$

with

$$\langle\omega_{kj}\rangle = 0 \quad . \tag{5.2.48}$$

We shall assume that these energy differences (which are those of nearest neighbour sites) are only correlated if they belong to the same bond, and uncorrelated otherwise. Thus

$$\langle\omega_{ki}\,\omega_{lj}\rangle = 2\,\langle\varepsilon_i^2\rangle\,(\delta_{kl}\delta_{ij} - \delta_{il}\delta_{kj}) \quad . \tag{5.2.49}$$

Here we define

$$\eta^2 = 2\,\langle\varepsilon_i^2\rangle \tag{5.2.50}$$

somewhat differently as compared to the last chapter.

If we insert $L_0\dot{n}_j$ into Σ_1^0 we observe that this correlation function describes a product mode $\omega_{lj}(c_l^\dagger c_j + c_j^\dagger c_l)$, where one factor is purely static and the other dynamic. The mode coupling approximation [Kawasaki, 1966], which has been introduced into the theory of Anderson localization by Götze [1978, 1979, 1981], consists in decoupling this correlation function into a product of two correlation functions:

$$\begin{aligned}&\left(\omega_{ki}(c_k^\dagger c_i + c_i^\dagger c_k)\,\Big|\,\frac{\mathrm{i}}{z-Q_1Q_0L}\,\Big|\,\omega_{lj}(c_l^\dagger c_j + c_j^\dagger c_l)\right)\\&\simeq \langle\omega_{ki}\,\omega_{lj}\rangle\left(c_k^\dagger c_i + c_i^\dagger c_k\,\Big|\,\frac{\mathrm{i}}{z-L}\,\Big|\,c_l^\dagger c_j + c_j^\dagger c_l\right) \quad .\end{aligned} \tag{5.2.51}$$

At the same time we neglect the projectors in front of L, which corresponds to a partial summation of the infinite series represented by the matrix element of $\mathrm{i}/(z-Q_1Q_0L)$. Approximations of this kind have been applied and tested in the theory of liquids and have been proved to be a powerful tool for going beyond perturbation theoretical approaches. Vollhardt and Wölfle [1980, see also Wölfle and Vollhardt, 1982] developed a non-perturbational diagrammatic theory for the Anderson localization problem. Their results agree with those of Götze's mode coupling theory. The latter, however, has the advantage of providing a more direct insight into the microscopic mechanism of localization.

With (5.2.51) we obtain for the first term, Σ_1^0, in (5.2.45)

$$\Sigma_{1ij}^0(z) = -J^2\eta^2 \left(c_i^\dagger c_j + c_j^\dagger c_i \,\middle|\, \frac{\mathrm{i}}{z-L} \,\middle|\, c_i^\dagger c_j + c_j^\dagger c_i\right) \tag{5.2.52}$$

if ij denote nearest neighbours. The diagonal elements obey

$$\Sigma_{1ii}^0(z) = -\sum_{j(\neq i)} \Sigma_{1ij}^0(z) \quad . \tag{5.2.53}$$

All other matrix elements are zero. Since this correlation function is a configurational average, all nearest neighbour elements are equal to each other. Consequently Σ_1^0 is proportional to $\psi(t=0)$ and we have

$$\{\Sigma_1^0(z)\,\psi^{-1}(t=0)\}_{ij} = M(z)\,\delta_{ij} \tag{5.2.54}$$

with some scalar function $M(z)$. Cross terms of (5.2.45) do not contribute within the decoupling approximation because of (5.2.48). The second term of (5.2.45) describes the correlation of $L_J\dot{n}_i$, i.e. the rate of change of the current in a scattering-free system. This correlation function does not contribute to the conductivity, since

$$L_J \sum_i \boldsymbol{R}_i \dot{n}_i = 0 \quad . \tag{5.2.55}$$

We will neglect the contribution of this term throughout, assuming that the scattering is dominated by the first term containing the correlation of the random potentials. This approximation corresponds to the hydrodymanic approximation for the density-density correlation function as applied by Belitz and Götze [1983]. So we have finally

$$\Sigma_{0ij}(z) = \frac{\mathrm{i}}{z + \mathrm{i}\,M(z)}\,\psi_{ij}(t=0) \tag{5.2.56}$$

and

$$\phi(z) = \frac{-1}{\mathrm{i}z + (\mathrm{i}z - M(z))^{-1}\,\psi(t=0)\,\phi^{-1}(t=0)}\,\phi(t=0) \quad . \tag{5.2.57}$$

Using (5.2.12) and (5.2.25) for the equal time correlation functions we obtain

$$\phi_{ij}(z) = \frac{\varrho^2}{\beta\gamma}(M(z) - \mathrm{i}z) \cdot G_{ij}(-6 + \mathrm{i}z\varrho(M(z) - \mathrm{i}z)/\gamma) \quad . \tag{5.2.58}$$

If $M(z)$ diverges like i/z in the dc limit, then also $\phi(z) \sim \mathrm{i}/z$, i.e. the density fluctuations are stationary, the system is localized. On the other hand, if $M_1 = M(0^+)$ is finite, we get in the dc limit

$$\phi_{ij}(z \to 0^+) = \frac{\varrho^2}{\beta\gamma} M_1 G_{ij}(-6) \quad . \tag{5.2.59}$$

Note that the dc limit of $\phi(z)$ is also given by the effective lattice Green's function taken at the lower band edge of the effective lattice Hamiltonian h. The matrix $G_{ij}(-6)$ is a real known matrix [Economou, 1983].

5.2.6 The Selfconsistency

Our task is now to determine $M(z)$. From (5.2.54),

$$\begin{aligned} M(z) = {} & 2\frac{\beta}{\gamma} J^2 \eta^2 \Big(c_1^\dagger c_2 \Big| \frac{\mathrm{i}}{z-L} \Big| c_2^\dagger c_1 \Big) \\ & + 2\frac{\beta}{\gamma} J^2 \eta^2 \Big(c_1^\dagger c_2 \Big| \frac{\mathrm{i}}{z-L} \Big| c_1^\dagger c_2 \Big) \quad , \end{aligned} \tag{5.2.60}$$

where 1,2 denote nearest neighbours. Due to the time reversal symmetry of the electronic Hamiltonian H_e we may interchange the indices of the spectral functions,

$$A_{ij} = A_{ji} \quad , \tag{5.2.61}$$

which leads to (c.f. (5.2.20))

$$M(z) = \alpha\beta\, \phi_{12}(z) + M_0(z) \quad , \tag{5.2.62}$$

where

$$M_0(z) = \alpha\beta \left(c_1^\dagger c_2 \Big| \frac{\mathrm{i}}{z-L} \Big| c_1^\dagger c_2 \right) \tag{5.2.63}$$

and

$$\alpha = 2 J^2 \eta^2 / \gamma \quad . \tag{5.2.64}$$

Together with (5.2.57) this gives a closed equation for $M(z)$. In particular the dc limit reads

$$M_1 = M_0(z = 0^+) + y\, M_1 \tag{5.2.65}$$

with

$$y = 2 \left(\frac{J\eta\varrho}{\gamma} \right)^2 \cdot 0.086 \quad , \tag{5.2.66}$$

where we have used [Economou, 1983]

$$G_{12}(-6) = 0.086 \quad . \tag{5.2.67}$$

This is solved immediately giving

$$M_1 = \frac{M_0(z=0^+)}{1-y} \quad . \tag{5.2.68}$$

We will approximate the non-critical contribution M_0 to $M(z)$ by the lowest order decoupling result of (5.2.20). Thus in the dc limit

$$M_0(0^+) = \alpha\pi\varrho^2 \quad . \tag{5.2.69}$$

5.2.7 The Conductivity

With (4.2.7), (5.1.15) and (5.2.36) we get for the conductivity

$$\sigma(z) = \left\langle -\frac{e^2\beta}{6\Omega} \sum_{ij} R_{ij}^2 \left\{ \frac{z}{z + \mathrm{i}\Sigma_0(z)\,\phi^{-1}(t=0)}\,\Sigma_0(z) \right\}_{ij} \right\rangle_{\text{conf}} \tag{5.2.70}$$

where for the moment we do not include the configurational average into the definition of the correlation functions. This expression is nothing else but a fancy way of writing down the Kubo formula

$$\sigma(z) = \frac{e^2\beta}{3\Omega} \left(\dot{\boldsymbol{X}} \left| \frac{\mathrm{i}}{z-L} \right| \dot{\boldsymbol{X}} \right) \quad , \tag{5.2.71}$$

where

$$\dot{\boldsymbol{X}} = \sum_i \boldsymbol{R}_i \dot{n}_i \quad . \tag{5.2.72}$$

Note the analogy to the formal solution of the linearized rate equation as presented in Sect. 4.5 (c.f. (4.5.10)). The collision matrix C appearing there is replaced by the frequency dependent density relaxation matrix $\Sigma_0(z)$. Indeed, for the more general case with electron-phonon interaction included the collision matrix C is an approximation to $\Sigma_0(z)$ in an appropriate hopping limit.

Equation (5.2.70) represents an infinite series of the form

$$\left\langle \sum_{ij} R_{ij}^2 \left(\Sigma_0 + \mathrm{i}\Sigma_0\,\phi^{-1}(t=0)\,\Sigma_0/z + \cdots\right)_{ij} \right\rangle \tag{5.2.73}$$

which describes quantum-mechanical tunneling processes from site to site, including all possible paths. The result has then to be averaged over all configurations. If, on the other hand, from the very beginning we include the configurational averaging into the definitions of all correlation functions, we find that all terms of the series vanish except the first one. This is seen if we employ isotropy and translational invariance of the averaged system and observe that due to particle conservation

$$\sum_j \Sigma_{0ij}(z) = \sum_i \Sigma_{0ij}(z) = 0 \quad . \tag{5.2.74}$$

Thus we have simply

$$\sigma(z) = -\frac{e^2\beta}{6\Omega}\sum_{ij} R_{ij}^2\,\psi_{ij}(z) = -\frac{e^2\beta}{6\Omega}\sum_{ij} R_{ij}^2\,\Sigma_{0ij}(z) \tag{5.2.75}$$

which is essentially Einstein's relation

$$\sigma(z) = \frac{e^2\varrho}{a^3}\,D(z) \tag{5.2.76}$$

between conductivity $\sigma(z)$ and generalized diffusion coefficient $D(z)$. Equation (5.2.75) shows that under the summation, Σ_0 can be replaced by the current-current correlation function ψ. With equation (5.2.56) for Σ_0 we obtain

$$\sigma(z) = -\frac{e^2\beta}{6\Omega}\sum_{ij} R_{ij}^2\,\frac{1}{-\mathrm{i}z + M(z)}\,\psi_{ij}(t=0) \tag{5.2.77}$$

and with (5.2.25)

$$\sigma(z) = \frac{e^2\gamma}{6\Omega}\,\frac{6\,Na^2}{M(z) - \mathrm{i}z} \quad . \tag{5.2.78}$$

Since the density of sites is $N/\Omega = a^{-3}$ we arrive at

$$\sigma(z) = \frac{e^2\,\gamma/a}{M(z) - \mathrm{i}z} \quad . \tag{5.2.79}$$

Using (5.2.68, 69) we have finally

$$\sigma(E) = \sigma(z = 0^+) = \frac{e^2}{a}\,\frac{\gamma}{\alpha\pi\varrho^2}\,(1 - y) \quad . \tag{5.2.80}$$

The coupling constant $y = y(E)$ contains the Fermi energy E (c.f. (5.2.66)). For $y(E) < 1$ the system is a conductor. The mobility edge E_C is given by $y(E = E_\mathrm{C}) = 1$. The separation from this critical point is $\tau = y - 1$, thus

$$\sigma \sim (-\tau)^\lambda \quad , \tag{5.2.81}$$

where the critical exponent $\lambda = 1$.

Alternatively, we can write with

$$\sigma_{00} = \frac{e^2}{a}\,\frac{\gamma}{\alpha\pi\varrho^2} \tag{5.2.82}$$

$$\sigma(E) = \sigma_{00} - 0.03\,\frac{e^2}{a} \tag{5.2.83}$$

since

$$\frac{y\gamma}{\alpha\pi\varrho^2} = \frac{G_{12}(-6)}{\pi} \simeq 0.03 \quad . \tag{5.2.84}$$

Note that within our approximations this number depends only on the underlaying lattice chosen and does not depend on any of the other microscopic parameters. The value of the nearest neighbour element of a Green's function at the lower band edge of a regular lattice with tight binding Hamiltonian h is not very much dependent on the lattice structure. We therefore expect that the value 0.03 in (5.2.84) is rather robust. It agrees numerically with the factor introduced by Mott [1972] in his discussion of the minimum metallic conductivity. Therefore we write

$$\sigma(E) = \sigma_{00} - \sigma_{\mathrm{M}} \tag{5.2.85}$$

$$\sigma_{\mathrm{M}} = 0.03\,\frac{e^2}{a} \quad . \tag{5.2.86}$$

The present theory, however, does not give a minimum metallic conductivity. The zeroth order conductivity σ_{00} is determined by the lowest order approximation to the conductivity and is non-zero everywhere in the band of the electronic Hamiltonian H_{e}. The full conductivity is then the difference between σ_{00} and σ_{M}, the conducting phase is characterized by $\sigma_{00} - \sigma_{\mathrm{M}} > 0$ without any finite jump. We have already discussed this result in more detail in the last chapter.

For vanishing disorder the correction σ_{M} to the zeroth order conductivity σ_{00} should depend on disorder and should vanish like the reciprocal mean free path. For strong disorder and close to the critical region, however, the mean free path approaches a and (5.2.85, 86) are valid [Economou et al., 1984].

5.3 The Potential Well Analogy

In order to make contact with the potential well analogy approach [Economou and Soukoulis, 1983; Economou et al., 1984, 1985; Papaconstantopoulos and Economou, 1981] to localization we first consider the mean square displacement and the generalized frequency dependent diffusion constant within the framework of our approach.

5.3.1 The Mean Square Displacement

The mean square displacement $\langle R^2(t)\rangle$ can be defined in terms of the density-density correlation function $\phi_{ij}(t)$, which has to be normalized by $\phi_{jj}(t=0) = \varrho/\beta$:

$$\langle R^2(t)\rangle = \frac{\beta}{N\varrho}\sum_{ij} R_{ij}^2\,\phi_{ij}(t) \quad . \tag{5.3.1}$$

Its Laplace transform is

$$\langle R^2(z)\rangle = \int_0^\infty \mathrm{e}^{izt}\,\langle R^2(t)\rangle\,dt \tag{5.3.2}$$

$$= \frac{\beta}{N\varrho}\sum_{ij} R_{ij}^2\,\phi_{ij}(z) \quad . \tag{5.3.3}$$

Using (4.2.7) and (5.1.15) we can write the conductivity as

$$\sigma(z) = -\frac{e^2\beta}{6\Omega} \sum_{ij} R_{ij}^2 \left(z^2\, \phi_{ij}(z) - \mathrm{i}z\, \phi_{ij}(t=0)\right) \tag{5.3.4}$$

and the mean square displacement can be expressed alternatively as

$$\langle R^2(z)\rangle = -\frac{6\,a^3}{e^2\,z^2\,\varrho}\,\sigma(z) \quad . \tag{5.3.5}$$

Using the generalized Einstein relation (5.2.76), this is equivalent to

$$\langle R^2(z)\rangle = -\frac{6}{z^2}\,D(z) \quad . \tag{5.3.6}$$

Alternatively, using the expression resulting from the mode coupling approximation (5.2.79), we have in terms of $M(z)$

$$\langle R^2(z)\rangle = -\frac{6\,a^2\,\gamma}{\varrho}\,\frac{1}{z^2\,(M(z)-\mathrm{i}z)} \quad . \tag{5.3.7}$$

Without going into the details of the calculation we give some results for the mean square displacement [Thomas and Weller, 1987]. The complex function $M(z)$ of complex argument z can be found by solving the set of selfconsistent equations (5.2.58) and (5.2.62). Then the Laplace back transform into the time domain can be performed. The long time asymptotics can be obtained analytically using the Tauberian theorems. It is found that $\langle R^2(t)\rangle \sim t^{\zeta}$ with $\zeta = 1$ in the conducting regime and $\zeta = 0$ in the insulating regime. The latter behaviour can be used to define the localization length ξ_{loc} treated in the next subsection. For critical disorder $\zeta = 2/3$. For short times the explicit Laplace back tranform yields $\zeta = 2$ in any case with a cross-over to $\zeta = 1$ (if the disorder is smaller than the critical one) at a certain time τ_{deph} , which can be identified [Thomas and Weller, 1989] with an intraband dephasing time [Kenkre and Reineker, 1982]. For $t < \tau_{\mathrm{deph}}$ the density fluctuation behaves like a quantum mechanical wave packet, where the mean square displacement moves with a group velocity given by the spectrum $E(\boldsymbol{k})$ of the effective ordered lattice Hamiltonian $\tilde{h}$. For $t > \tau_{\mathrm{deph}}$ scattering becomes so effective that the propagation becomes diffusive.

5.3.2 The Localization Length

In the localized regime $M(z)$ diverges like i/z for $z \to 0$, which implies the same divergence for $\langle R^2(z)\rangle$. Therefore the long time limit of $\langle R^2(t)\rangle$ is constant. It is identified with the square of the localization length ξ_{loc} ,

$$\xi_{\mathrm{loc}}^2 = \lim_{t\to\infty} \langle R^2(t)\rangle \quad . \tag{5.3.8}$$

Thus in terms of the Laplace tranforms

$$\xi_{\rm loc}^2 = \lim_{z\to 0} (-{\rm i}z) \langle R^2(z) \rangle \tag{5.3.9}$$

$$= \frac{6\,a^2\,\gamma}{\varrho} \lim_{z\to 0} \frac{1}{(-{\rm i}z)\,M(z)} \tag{5.3.10}$$

or alternatively

$$\xi_{\rm loc}^2 = 6 \lim_{z\to 0} D(z)/(-{\rm i}z) \quad . \tag{5.3.11}$$

We can also relate $\xi_{\rm loc}$ to the static dielectric susceptibility $\tilde{\chi}_0 = \tilde{\chi}(z=0)$. The frequency dependent dielectric susceptibility $\tilde{\chi}(z)$ is related to the conductivity by

$$\tilde{\chi}(z) = {\rm i}\,\sigma(z)/z \quad . \tag{5.3.12}$$

Thus we get

$$\xi_{\rm loc}^2 = \frac{6\,a^3}{e^2\,\varrho}\,\tilde{\chi}_0 \quad . \tag{5.3.13}$$

The critical behaviour of $\xi_{\rm loc}(E)$ when approaching $E_{\rm C}$ from below follows from (5.2.58) and (5.2.62), which can be combined to give

$$1 = \delta\,G_{12}(-6 - 6a^2/\xi_{\rm loc}^2)/G_{12}(-6) \quad . \tag{5.3.14}$$

From the analytic properties of the effective lattice Green's function $G_{12}(E)$ for E below the lower band edge of the effective ordered Hamiltonian h [Economou, 1983] we find that in three dimensions

$$\xi_{\rm loc} \sim \tau^{-\lambda} \tag{5.3.15}$$

with the same exponent $\lambda = 1$ which characterizes the critical behaviour of the conductivity above $E_{\rm C}$. The static susceptibility $\tilde{\chi}_0$ then diverges like $\tau^{-2\lambda}$. These results are well known in the theory of Anderson localization (c.f.Sect. 2.4.2).

5.3.3 More Realistic Model Hamiltonians

The analogy of (5.3.14) to the solution of the potential well problem has been realized by Economou and Soukoulis [1983]. Consider the effective ordered lattice Hamiltonian h. The corresponding Green's function is

$$G(E) = \frac{1}{h - E} \quad . \tag{5.3.16}$$

If at site 1 we introduce a potential $\varepsilon_1 < 0$ there will be a bound state at an energy

$$E = -6 - E_{\rm b} \tag{5.3.17}$$

in three dimensions if $-\varepsilon_1$ is sufficiently large. Since $E = -6$ is the lower undisturbed band edge of h, $E_{\rm b}$ is the binding energy of the bound state. It is determined by the pole of

$$\frac{1}{h + \varepsilon_1 - E} \tag{5.3.18}$$

which is given by the solution of

$$1 = -\varepsilon_1 G_{11}(-6 - E_b) \tag{5.3.19}$$

if

$$-\frac{1}{\varepsilon_1} \leq G_{11}(-6) \quad . \tag{5.3.20}$$

Formally this equation looks similar to (5.3.14) if we identify the binding energy E_b with $6a^2/\xi_{\mathrm{loc}}^2$. To complete the analogy one can easily relate the nearest neighbour matrix element G_{12} in (5.3.14) to the diagonal matrix element G_{11}.

The potential ε_1 is then given in terms of the coupling constant y, which is proportional to the second moment of the site energy distribution function. For the simple Anderson model treated here y is a given model parameter. If one tries to model an amorphous semiconductor using a more realistic Hamiltonian, containing parameters describing Si-Si and Si-H bonds etc. [Papaconstantopoulos and Economou, 1980; Economou et al., 1985; Zdetsis et al., 1985a; Zdetsis, 1987], the coupling constant y is not a priori given. It is then advantageous to use (5.2.83) which relates y to σ_{00}, the zeroth order decoupling approximation to the conductivity, $y = \sigma_M/\sigma_{00}$. The approximation σ_{00} can then be calculated by, e.g., the coherent potential approximation on the basis of the more realistic Hamiltonian. Then y is determined and the potential ε_1 is found. The binding energy determines the localization length as a function of energy. On the conducting side of E_C one can immediately use e.g. (5.2.85) to determine the conductivity. This procedure is, however, only approximative, because it requires that $y\sigma_{00} = \sigma_M$, which holds only if $M_0(0^+)$, and thus σ_{00}, is given by the decoupling result(5.2.69).

It is clear that, whatever approximation we use to determine σ_{00}, a relation comparable to (4.3.8) must follow, which ultimately determines the prefactor of the temperature dependent conductivity of the semiconductor. A value of $\sigma_0 = 137\Omega^{-1}\,\mathrm{cm}^{-1}$ has been found [Zdetsis, 1986] from the potential well analogy approach in close agreement with our result given in Sect. 4.8. Moreover, results for the critical disorder and the localization length agree with those of other theories and numerical calculations [Zdetsis et al., 1985b].

The modifications introduced by the electron-phonon coupling have not yet been studied within the potential well analogy approach. Following the lines of the next section a generalization could certainly be achieved. We consider it, however, sufficient to start from the simple Anderson Hamiltonian instead of treating more realistic Hamiltonians. Within the potential well analogy approach the latter form the basis for a calculation of the zeroth order conductivity $\sigma_{00}(E)$. This function turns out to be rather featureless close to the lower conduction band edge. As far as the dc conductivity is concerned the Anderson Hamiltonian will produce comparable results if the model parameters are chosen as described in the last chapter.

5.4 The Interacting Case

So far we have not found any new results. We have merely reformulated Götze's and Belitz and Götze's theory in a representation more suitable to a generalization for the interacting case, where hopping processes have to be described. At the same time we have introduced additional assumptions in order to simplify some of the expressions. In particular, the selfconsistency equations became quite trivial due to the neglection of additional terms in $\Sigma_1(z)$ as discussed below (5.2.55). For a more thorough discussion of Anderson localization see the work of Götze and coworkers. In this Section we add H_p and H_ep to the electronic Hamiltonian H_e. We show that this interaction influences the resultant differential conductivity in a characteristic way.

5.4.1 The Canonical Transformation

Remember that the theory has to be generalized in such a way as to describe hopping processes in the appropriate limit. In order to calculate the hopping rates in conventional hopping theory one takes the interaction Hamiltonian H_ep between electronic eigenstates calculated for an isolated pair of sites [Miller and Abrahams, 1960]. It is then clear that this approach can not describe such situations where inelastic processes act on carriers in *extended states*. In a more general theory the coupling of all sites has to be retained. In other words, we have to go beyond the binary limit of conventional hopping theory. The hop rates should be calculated taking into account the full quantum mechanically coupled system of sites. In principle, one could diagonalize the electronic Hamiltonian H_e and take the interaction between the eigenstates. This procedure, however, is not feasible in the disordered case.

Fortunately, there is a way out of this difficulty. At least in the binary limit of conventional hopping theory one can avoid the diagonalization of H_e. A canonical transformation can be applied to the full Hamiltonian. Then the density relaxation rate is calculated for the binary case to lowest order in J. In the limit of single phonon processes it turns out that the dc limit of this density relaxation rate is identical to the collision matrix C given by the hoprates of conventional hopping theory [Dersch et al., 1983]. Retaining the phonon processes at all orders, the small polaron multi-phonon hoprate results. We will apply this transformation and calculate the density relaxation rate approximately, however, beyond the binary limit. We are then sure that in the appropriate limit of pure hopping transport we obtain the known results of conventional hopping theory. On the other hand, this approach ensures that the localization phenomenon is still contained in the theory if the phonons are switched off.

The canonical transformation mentioned is that applied in small polaron theory [Holstein, 1959; Schnakenberg, 1966]. Here we just quote the result. The transformed Hamiltonian reads (neglecting additional terms describing interacting quasiparticles)

$$\tilde{H} = \sum_i (\tilde{\varepsilon}_i - E) n_i + J \sum_{ij} c_i^\dagger c_j \, \mathrm{e}^{s_{ij}} + H_\mathrm{p} \tag{5.4.1}$$

where H_p is that of (4.1.5). The single site energies are renormalized due to a relaxation of the deformable lattice around an occupied site. The transfer matrix elements are renormalized by a factor $\exp(s_{ij})$ containing the phonon degrees of freedom:

$$s_{ij} = \sum_q \frac{A_{ij}(q)\, b_q^\dagger - A_{ij}(-q)\, b_q}{\omega_q} \quad . \tag{5.4.2}$$

$A_{ij}(q)$ is a new effective electron-phonon coupling constant (the difference $A_{iq} - A_{jq}$). Obviously the exponential $\exp(s_{ij})$ describes the multi-phonon processes contained in the transfer of the new quasiparticle. Eventually, however, we will restrict ourselves to single-phonon processes. We will, therefore, also neglect the renormalization of the single site energies assuming that the binding energy $\tilde{\varepsilon}_i - \varepsilon_i$ is small compared to typical disorder energies given by η. (In the following we will omit the $\sim$ on the top of H). So our new Hamiltonian differs from (4.1.7) by the absence of an extra interaction term H_ep and by a dynamical transfer matrix $J \exp(s_{ij})$ for nearest neighbours. This transfer matrix describes the dynamically fluctuating coupling between nearest neighbours giving rise to incoherent inelastic transport processes, once the thermodynamic average over the phonon system is taken. Our new Hamiltonian is the quantum analog of the Haken–Strobel Hamiltonian [Haken and Strobel, 1967] describing classically fluctuating coupling between sites. Actually, in the following we will not be interested in the quantum nature of the phonons, as we will take the high temperature limit for the phonon system. The reasons for using the quantum mechanical description are, first, to make contact with conventional hopping theory, and second, the fact that the Hamiltonian is independent of time.

5.4.2 Some Results from Small Polaron Theory

We will now quote some well known results of small polaron theory which can be found in textbooks (Mahan [1983], see also [Holstein, 1959; Emin, 1974]). The thermal average of $\exp(s_{ij})$ is called the band narrowing factor. It is smaller than one and decreases with increasing temperature. It may be written as

$$\mathrm{e}^{-\theta/2} = \langle \mathrm{e}^{s_{ij}} \rangle \tag{5.4.3}$$

where

$$\theta = \sum_q \left| \frac{A_q}{\omega_q} \right|^2 \coth \frac{\beta \omega_q}{2} > 0 \quad . \tag{5.4.4}$$

We will neglect in the following the dependence of θ on the particular pair of sites, writing A_q for the coupling constant in (5.4.4). In the high-temperature limit we have

$$\theta \cong 2 \sum_{\boldsymbol{q}} \left|\frac{A_{\boldsymbol{q}}}{\omega_{\boldsymbol{q}}}\right|^2 (\beta\omega_{\boldsymbol{q}})^{-1} \tag{5.4.5}$$

which is porportional to kT.

When describing hopping processes we will also encounter a phonon correlation function defined by

$$\Lambda(t + \mathrm{i}\beta/2) = \langle \{\mathrm{e}^{s_{ij}+\theta/2} - 1\} \{\mathrm{e}^{s_{kl}(-t)+\theta/2} - 1\} \rangle \quad . \tag{5.4.6}$$

Performing the average over the equilibrium phonon system we get

$$\Lambda(t) = \mathrm{e}^{\varphi(t)} - 1 \tag{5.4.7}$$

with

$$\varphi(t) = \sum_{\boldsymbol{q}} \left|\frac{A_{\boldsymbol{q}}}{\omega_{\boldsymbol{q}}}\right|^2 \operatorname{cosech}\frac{\beta\omega_{\boldsymbol{q}}}{2} \cos(\omega_{\boldsymbol{q}} t) \tag{5.4.8}$$

$$\cong 2 \sum_{\boldsymbol{q}} \left|\frac{A_{\boldsymbol{q}}}{\omega_{\boldsymbol{q}}}\right|^2 (\beta\omega_{\boldsymbol{q}})^{-1} \cos(\omega_{\boldsymbol{q}} t) \tag{5.4.9}$$

where we have once more applied the high temperature limit.

5.4.3 The Generalized Collision Matrix

We are now in the position to calculate the density relaxation rate $\Sigma_0(z)$ for the interacting system. Following Schnakenberg [1968] we make the exact decomposition

$$\dot{n}_i = \dot{n}_i^{\mathrm{t}} + \dot{n}_i^{\mathrm{h}} \tag{5.4.10}$$

where

$$\dot{n}_i^{\mathrm{t}} = \mathrm{i}J_{\mathrm{T}} \sum_j c_j^\dagger c_i + \mathrm{HC} \tag{5.4.11}$$

and

$$\dot{n}_i^{\mathrm{h}} = \mathrm{i}J_{\mathrm{T}} \sum_j (\mathrm{e}^{s_{ji}+\theta/2} - 1) c_j^\dagger c_i + \mathrm{HC} \quad . \tag{5.4.12}$$

Here

$$J_{\mathrm{T}} = J\,\mathrm{e}^{-\theta/2} \tag{5.4.13}$$

is the (slightly) reduced thermodynamically averaged transfer matrix element.

Accordingly $\Sigma_0(z)$ is decomposed into four terms:

$$\Sigma_{0ij}^{\mathrm{t}}(z) = \left(\dot{n}_i^{\mathrm{t}} \left| \frac{\mathrm{i}}{z - Q_0 L} \right| \dot{n}_j^{\mathrm{t}}\right) \tag{5.4.14}$$

describing density relaxation due to tunneling processes, and

$$\Sigma^{\mathrm{h}}_{0ij}(z) = \left(\dot{n}^{\mathrm{h}}_i \,\middle|\, \frac{\mathrm{i}}{z - Q_0 L} \,\middle|\, \dot{n}^{\mathrm{h}}_j\right) \tag{5.4.15}$$

describing density relaxation due to hopping processes. In addition we have two cross terms.

Let us now concentrate on Σ^{h}_0. We first show that this matrix reduces to the collision matrix C in the binary hopping limit. To this end we do not include the configurational average into the definition of the correlation functions. In contrast to the Kubo correlation function the Van Hove correlation function

$$S^{\mathrm{h}}_{ij}(t) = \langle \dot{n}^{\mathrm{h}}_i \, \dot{n}^{\mathrm{h}}_j(-t) \rangle \tag{5.4.16}$$

has the advantage to decouple automatically in the binary limit if we calculate it to order J^2 [Belitz and Schirmacher, 1983]. We insert $\dot{n}^{\mathrm{h}}_i$ into S^{h} and get terms like

$$J^2_{\mathrm{T}} \sum_{kl} \left\langle (\mathrm{e}^{s_{ik}+\theta/2} - 1)\, c^\dagger_i c_k \, (\mathrm{e}^{s_{lj}(-t)+\theta/2} - 1)\, c^\dagger_l(-t)\, c_j(-t) \right\rangle \quad . \tag{5.4.17}$$

To lowest order in J (i.e. keeping only the diagonal part of the electronic operators) this decouples into

$$J^2_{\mathrm{T}} \, \Lambda(t + \mathrm{i}\beta/2) \sum_{kl} \langle c^\dagger_i c_k \, c^\dagger_l(-t)\, c_j(-t) \rangle \quad . \tag{5.4.18}$$

This leads to

$$S^{\mathrm{h}}_{ij}(t) = \Lambda(t + \mathrm{i}\beta/2)\, S^{\mathrm{t}}_{ij}(t) \tag{5.4.19}$$

where

$$S^{\mathrm{t}}_{ij}(t) = \langle \dot{n}^{\mathrm{t}}_i \, \dot{n}^{\mathrm{t}}_j(-t) \rangle \tag{5.4.20}$$

is the tunneling $\dot{n}$–$\dot{n}$ correlation function (to lowest order in J here). It is easily calculated for the binary case giving

$$S^{\mathrm{t}}_{ij}(t) = -J^2_{\mathrm{T}} \, f^0_j (1 - f^0_i)\, \mathrm{e}^{-\mathrm{i}\omega_{ij} t} + (i \rightleftharpoons j) \quad . \tag{5.4.21}$$

Using (5.4.21) and (5.4.19), taking the Fourier transform, then using (5.1.5, 9, 13) we get, for i, j nearest neighbours

$$\Sigma^{\mathrm{h}}_{0ij}(z = 0^+) = -\pi \, J^2_{\mathrm{T}} \, f^0_j (1 - f^0_i)\, \mathrm{e}^{-\beta\omega_{ij}/2}\, \Lambda(\omega_{ij}) + (i \rightleftharpoons j) \tag{5.4.22}$$

with

$$\Lambda(\omega) = \frac{1}{2\pi} \int_{-\infty}^{\infty} \mathrm{e}^{\mathrm{i}\omega t}\, \Lambda(t)\, dt \quad . \tag{5.4.23}$$

This is the multi-phonon hopping rate used in small polaron theory [Holstein, 1959; Emin, 1974]. Taking the phonon propagator to lowest order in $A_{\boldsymbol{q}}$, i.e.

neglecting multi-phonon processes,

$$\Lambda(t) = \varphi(t) \quad , \tag{5.4.24}$$

and we recover [Dersch et al., 1983] the hopping rates of the conventional Miller–Abrahams theory. It is also seen that in the binary limit the tunneling contribution to the density relaxation rate vanishes in the dc limit since ω_{ij} is non-zero and the Fourier transform of S^{t} for $\omega \to 0$ vanishes. We can now insert (5.4.22) for Σ_0^{h} into (5.2.75) for the conductivity and arrive at the formal solution of the linearized rate equation as given in (4.5.10).

Let us now go back to (5.4.15) and consider the hopping contribution to the density relaxation beyond lowest order in J, taking into account the full quantum mechanically coupled system of sites. The corresponding van Hove correlation function $S^{\mathrm{h}}(t)$ will then no longer decouple into an electron and a phonon factor. We will, however, apply a dynamical mode-coupling approximation by decoupling $S^{\mathrm{h}}(t)$ in an analogous way. Following the same procedure as applied in the binary limit, the result is

$$\Sigma_{0ij}^{\mathrm{h}}(z) = \frac{1}{2\pi \mathrm{i}} \int_{-\infty}^{\infty} d\omega \int_{-\infty}^{\infty} d\omega' \, \frac{1}{\omega - z} \, \Lambda(\omega - \omega') \, \Sigma_{0ij}^{\mathrm{t}''}(\omega') \tag{5.4.25}$$

where we have already taken the high T limit, i.e. $\sinh(\beta\omega/2) \cong \beta\omega/2$.

In deriving this expression we have been somewhat careless with the projector Q_0 in front of L (see (5.4.15)), which also should contain phonon contributions. Following the discussion below (5.2.73), however, it can be seen that the additional terms generated by Q_0 do not contribute to the conductivity if the configurational average is included into the correlation functions. In a $\boldsymbol{k}$-space formulation of the theory the conductivity is given by the long wavelength limit of the correlation function. It can be shown that in this limit Q_0 can be replaced by 1. This will also be done in Σ_0^{h} after the dynamical mode-coupling approximation has been performed.

A particularly simple form of Σ_0^{h} is gained if we assume the most simple time dependence of the phonon correlation function, namely an exponential one

$$\Lambda(t) \sim \mathrm{e}^{-\omega_0 t} \quad , \tag{5.4.26}$$

where ω_0 is a typical phonon frequency [Jonson and Girvin, 1979; Girvin and Jonson, 1980]. Then

$$\Lambda(\omega) = \bar{\Lambda} \, \frac{\omega_0/\pi}{\omega^2 + \omega_0^2} \tag{5.4.27}$$

with

$$\bar{\Lambda} = \int_{-\infty}^{\infty} \Lambda(\omega) \, d\omega \quad . \tag{5.4.28}$$

To lowest order in the electron phonon coupling $A_{\boldsymbol{q}}$ and in the high T limit this is given by

$$\bar{\Lambda} = \theta \tag{5.4.29}$$

where θ is given by (5.4.4). The dc limit of Σ_0^{h} reads simply

$$\Sigma_{0ij}^{\mathrm{h}}(z=0^+) = \theta\,\Sigma_{0ij}^{\mathrm{t}}(z=\mathrm{i}\omega_0) \quad . \tag{5.4.30}$$

We will further simplify this expression by neglecting phonon contributions in Σ_0^{t} (5.4.14). This approximation can be made plausible by inspection of (5.4.25) which shows that Σ_0^{h} is a convolution of a phonon propagator and the spectrum of the tunneling density relaxation rate. The result of this convolution may be quantitatively changed by neglecting phonon contributions to the spectrum. Its qualitative physical influence on the total density relaxation should, however, still be described reasonably well. The essential point is an additional density relaxation channel which at finite temperature is active for all positions of the Fermi energy. We will now study the effect of this relaxation channel on the conductivity.

We first note, that the two cross terms appearing in $\Sigma_0(z)$ (c.f. (5.4.14, 15)) vanish after the dynamical mode coupling approximation has been performed. According to (5.2.75) the total conductivity then has the form of two parallel conducting paths

$$\sigma = \sigma_{\mathrm{t}} + \sigma_{\mathrm{h}} \tag{5.4.31}$$

where σ_{t} contains the tunneling rate Σ_0^{t} and σ_{h} the hopping rate Σ_0^{h}. Note, however, that these two contributions are not independent of each other (like, e.g., in (2.8.16)), since σ_{h} is determined by σ_{t}, as can be seen from (5.4.25). On the other hand, σ_{t} itself will be shown to be influenced considerably by the hopping processes. This new effect is called phonon induced delocalization.

5.4.4 Current Relaxation Rates

Since we have neglected phonon contributions to $\Sigma_0^{\mathrm{t}}(\mathrm{i}\omega_0)$ we can solve the self-consistency equation for $M(z=\mathrm{i}\omega_0) = M_2$ first:

$$M_2 = M_0(\mathrm{i}\omega_0) + \alpha\beta\,\phi_{12}(\mathrm{i}\omega_0)|_{\text{no phonons}} \quad . \tag{5.4.32}$$

As a slight complication we now have to consider the effective lattice Green's function G not at the lower band edge, but outside the band at an energy

$$E_2 = -6 - \omega_0\,\varrho(M_2+\omega_0)/\gamma \quad . \tag{5.4.33}$$

At this energy the nearest neighbour element of G is real and a known function of M_2. The density-density correlation function at $\mathrm{i}\omega_0$ is then given by (c.f. (5.2.58))

$$\phi_{12}(\mathrm{i}\omega_0)|_{\text{no phonons}} = \frac{\varrho^2}{\beta\gamma}\,(\omega_0+M_2)\,G(E_2(M_2)) \quad . \tag{5.4.34}$$

The first term in (5.4.32) will be approximated by

$$M_0(\mathrm{i}\omega_0) \cong M_0(0^+) \quad . \tag{5.4.35}$$

This is justified since

$$\left(c_1^\dagger c_2 \,\middle|\, \frac{\mathrm{i}}{z-L} \,\middle|\, c_1^\dagger c_2\right) \tag{5.4.36}$$

varies on an energy scale determined by electronic energies which are typically larger than phonon energies ω_0.

The selfconsistency equations can be solved numerically yielding M_2, which is inserted into

$$M_1 = M(0^+) = M_0(0^+) + \alpha\beta\,\phi_{12}(0^+) \tag{5.4.37}$$

where $\phi_{12}(0^+)$ now contains two density relaxation kernels. It is given by a generalization of (5.2.36) which in the dc limit reads

$$\phi(0^+) = \frac{\mathrm{i}}{z + \mathrm{i}\Big(\Sigma_0^{\mathrm{t}}(0^+) + \theta\,\Sigma_0^{\mathrm{t}}(\mathrm{i}\omega_0)\Big)\,\phi^{-1}(t=0)}\,\phi(t=0) \quad . \tag{5.4.38}$$

Its nearest neighbour matrix element is given by the nearest neighbour matrix element of the effective lattice Green's function G at the energy -6, i.e. at the lower band edge, and we obtain

$$\phi_{12}(0^+) = \frac{\varrho^2}{\beta\gamma}\,G_{12}(-6)\left(\frac{1}{M_1} + \frac{\theta}{\omega_0 + M_2}\right)^{-1} \tag{5.4.39}$$

which leads to a selfconsistency equation for M_1 using (5.4.37):

$$M_1 = M_0 + y\left(\frac{1}{M_1} + \frac{\theta}{\omega_0 + M_2}\right)^{-1} \quad . \tag{5.4.40}$$

This is the generalization of (5.2.65), and reduces to it in the limit $\theta = 0$. We have once again used the argument below (5.4.35) and have neglected phonon contributions to the equal time correlation functions in (5.4.38).

5.4.5 The Conductivity

We are now ready to discuss the dc conductivity $\sigma(E)$. It is given in terms of M_1 and M_2 as

$$\sigma = \sigma_{\mathrm{t}} + \sigma_{\mathrm{h}} = \frac{e^2}{a}\frac{\gamma}{M_1} + \frac{e^2}{a}\,\theta\,\frac{\gamma}{\omega_0 + M_2} \quad . \tag{5.4.41}$$

From (5.4.40) we can now see that the hopping contribution (given by M_2) to the conductivity influences the tunneling contribution (given by M_1). This is the phonon induced delocalization.

The interrelation between σ_{t} and σ_{h} can be discussed in a more transparent way by rewriting (5.4.41) in terms of the conductivities and observing that (c.f. (5.2.84))

$$y\,\sigma_{00} = \sigma_{\rm M} \tag{5.4.42}$$

is a constant (i.e. independent of the energy). Then the selfconsistency equation (5.4.40) for M_1 reads

$$\sigma_{\rm t}^{-1} = \sigma_{00}^{-1} + \frac{y}{\sigma_{\rm t} + \sigma_{\rm h}} \tag{5.4.43}$$

or, equivalently,

$$\sigma = \sigma_{00} + \sigma_{\rm h} - \sigma_{\rm M} + \frac{\sigma_{\rm h}}{\sigma}\sigma_{\rm M} \quad . \tag{5.4.44}$$

The hopping conductivity is a given function determined by M_2 from (5.4.32, 34). Note that $\sigma_{\rm h} \ll \sigma_{\rm M}$. This is indeed the case as can be seen by using the phonon free result (4.3.9) for the moment. Then $\sigma(E) = \sigma_{\rm M}$ for an energy $E = 2E_{\rm C}$, which is well above $E_{\rm C}$. The hopping conductivity, on the other hand, is typically much less than the tunneling conductivity above $E_{\rm C}$.

Equation (5.4.44) is solved by

$$\sigma = \tfrac{1}{2}(\sigma_{00} + \sigma_{\rm h} - \sigma_{\rm M}) + \tfrac{1}{2}\{(\sigma_{\rm h} + \sigma_{00} - \sigma_{\rm M})^2 + 4\,\sigma_{\rm h}\,\sigma_{\rm M}\}^{1/2} \tag{5.4.45}$$

where only the positive sign in front of the square root is allowed physically. The mobility edge is determined by $\sigma_{00} = \sigma_{\rm M}$. For $\theta = 0$ we recover (5.2.85).

Let us consider first the case for $E > E_{\rm C}$, i.e. $\sigma_{00} > \sigma_{\rm M}$. Then

$$\sigma \cong \sigma_{00} - \sigma_{\rm M} + \frac{\sigma_{\rm h} + \sigma_{00}}{\sigma_{\rm h} + \sigma_{00} - \sigma_{\rm M}}\sigma_{\rm h} \tag{5.4.46}$$

and the conductivity is given by the phonon free result $\sigma_{00} - \sigma_{\rm M}$ plus a small hopping contribution roughly proportional to $\sigma_{\rm h}$.

On the other hand, for $\sigma_{00} + \sigma_{\rm h} < \sigma_{\rm M}$, i.e. $E < E_{\rm C}$, we have

$$\begin{aligned}\sigma &\cong \tfrac{1}{2}(\sigma_{00} + \sigma_{\rm h} - \sigma_{\rm M}) + \tfrac{1}{2}\,|\sigma_{\rm h} + \sigma_{00} - \sigma_{\rm M}| \\ &\quad\times\left(1 + \frac{2\,\sigma_{\rm h}\sigma_{\rm M}}{\sigma_{\rm h} + \sigma_{00} - \sigma_{\rm M}}\right) = \frac{\sigma_{\rm M}}{\sigma_{\rm M} - \sigma_{00} - \sigma_{\rm h}}\sigma_{\rm h}\end{aligned} \tag{5.4.47}$$

and σ is given roughly by the pure hopping conductivity $\sigma_{\rm h}$ alone.

Close to $E_{\rm C}$, and even below $E_{\rm C}$, there is a contribution from both $\sigma_{\rm h}$ and $\sigma_{\rm t}$. Let us consider a special case for $E < E_{\rm C}$, such that $\sigma_{00} + \sigma_{\rm h} = \sigma_{\rm M}$. Then

$$\sigma = \sigma_{\rm t} + \sigma_{\rm h} = (\sigma_{\rm h}\,\sigma_{\rm M})^{1/2} \tag{5.4.48}$$

and the tunneling contribution is given by

$$\sigma_{\rm t} = \sigma_{\rm h} \cdot \{(\sigma_{\rm M}/\sigma_{\rm h})^{1/2} - 1\} \tag{5.4.49}$$

which can be much larger than $\sigma_{\rm h}$ at this energy. So even below $E_{\rm C}$ the conductivity is dominated by tunneling processes induced by the incoherent inelastic hopping processes. These give rise to the characteristic gradual decrease of $\sigma(E)$ for E below, but close to, $E_{\rm C}$.

5.4.6 The Mean Square Displacement

Following Sect. 5.3.1 the mean square displacement can also be studied in the interacting case. The following results are found [Thomas and Weller, 1989]: As in the interaction-free case at very short times, $t < \omega_0^{-1}$, $\langle R^2(t) \rangle \sim t^2$, with the same group velocity. Phonons act at this time scale as an additional source of static disorder, which is not yet felt by the extending wave packet.

At long times, $t \to \infty$, there is always diffusion and no localization occurs. The dc-diffusion constant exceeds that of the interaction-free case. Unfortunately we have not yet been able to obtain the Laplace back-transform for times of the order of ω_0^{-1} to see the onset of the dynamics introduced by the phonons.

For vanishing disorder one should consider intraband dephasing due to electron-phonon scattering. In our present approach we have neglected this mechanism and, therefore, are restricted to the case of large disorder.

6. The Long-Ranged Random Potential

Thus far we have treated amorphous semiconductors as isotropic and homogeneous systems. Any disorder is of course extremely anisotropic and inhomogeneous on a microscopic scale of the order of interatomic nearest neighbour distances. The length scale relevant for transport, however, can be considered large compared to the length scale inherent in the disorder which allows for our neglect of any nonuniformity.

In this chapter we shall investigate the inverse problem, namely the presence of a long-ranged disorder potential. The length scale of this potential will be some 500 Å and thereby exceed the inelastic scattering length, L_{i}. We shall, therefore, treat the long-ranged potential in the semiclassical approximation: its influence is described by a modulation of the electronic energies.

Long-ranged random potentials of electrostatic nature have been discussed for heavily doped crystalline semiconductors by Bonch–Bruevich [1962, 1970], by Kane [1963], and by Keldysh and Proshko [1963]. The idea that the presence of a long-ranged random potential of essentially electrostatic nature would be present in amorphous semiconductors was first suggested by Tauc [1970] in his interpretation of the exponential absorption edges. Fritzsche [1971] used this concept in order to explain the discrepancy between a rather large density of midgap states as suggested by the high values of the ac conductivity in unhydrogenated amorphous semiconductors and the rather low optical absorption at lower photon energies. Last and Thouless [1971], Kirkpatrick [1971], and Stinchcombe [1973] have employed percolation theory in order to calculate the conductivity for a classical model of a random potential. Independently, Shklovskii and Efros [1970, 1971, 1972, see also 1984] derived a long-ranged random potential in their theory of the transport properties of heavily doped and highly compensated crystals. They realized that amorphous semiconductors in a way represented the case of a perfectly compensated semiconductor. Bonch–Bruevich [1970] and his school discussed random fields of different origins and lengthscales in amorphous semiconductors and studied the influence of these fields on various properties [see, e.g., Bonch–Bruevich, 1983]. Overhof and Beyer [1981; see also Beyer and Overhof, 1984] postulated such a potential in order to explain the "activation energy" E_Q^* observed in their study of the Q function.

In this chapter we shall start with the discussion of the electrostatic potential in doped compensated crystals following the treatment given by Jäckle [1980] which can be adapted most easily to the case of amorphous semiconductors. This will lead us to the conclusion that a significant electrostatic long-ranged potential must be present in doped amorphous semiconductors, in particular if high-quality material is investigated. There are other sources of a long-ranged random potential which will also be described.

We shall then proceed to explore the influence of such a potential on the various transport properties. For the ease of discussion we shall use the simple Standard Transport Model of Sect. 2.6 instead of the more realistic results of Sect. 4.10. The influence of a random potential has been studied by two different numerical methods, a resistor model and a random walk model. Both methods are presented in some detail and the resulting influence on the transport properties is exemplified.

6.1 The Random Potential of a Heavily Doped Compensated Crystal

Following a paper of Jäckle [1980] we shall treat the case of an n-type crystal with a rather high density of $\bar{n}_A$ acceptors and $\bar{n}_D$ donors. Since the donor density exceeds the acceptor density the acceptor levels will be filled (at sufficiently low temperatures) and thus be negatively charged. Most of the donors will be ionized and positively charged but some will be occupied by one electron and hence be neutral. In the following we shall treat the positive charge of the donors as a uniformly distributed background charge and concentrate on the charged acceptors. The latter are supposed to be distributed randomly in space and, therefore, the correlation function for the acceptor density fluctuation

$$\Delta n_A(\boldsymbol{r}) = n_A(\boldsymbol{r}) - \bar{n}_A \tag{6.1.1}$$

is given by

$$\overline{\Delta n_A(\boldsymbol{r}) \cdot \Delta n_A(\boldsymbol{r}')} = \bar{n}_A \cdot \delta(\boldsymbol{r} - \boldsymbol{r}') \quad . \tag{6.1.2}$$

From this expression we obtain the correlation function for the resulting electrostatic random potential (in a medium with a dielectric constant ε_s)

$$\overline{V(\boldsymbol{r}) \cdot V(\boldsymbol{r}')} = \int \frac{d^3k}{(2\pi)^3} \, \bar{n}_A \, v^2(\boldsymbol{k}) \cdot e^{i\boldsymbol{k}(\boldsymbol{r}-\boldsymbol{r}')} \tag{6.1.3}$$

with

$$v(\boldsymbol{k}) = \frac{e^2}{\varepsilon_0 \, \varepsilon_s \, k^2} \quad . \tag{6.1.4}$$

The integral diverges as $\boldsymbol{k}$ approaches zero independent of the value of $(\boldsymbol{r} - \boldsymbol{r}')$ as was already noted by Keldysh and Proshko [1963]. This divergency still persists if we consider the coarse-grained density fluctuation with typical length l instead

$$\langle \Delta n_A(\boldsymbol{r}) \rangle_l = \int d^3r' \, \Delta n_A(\boldsymbol{r}') \, g_l(\boldsymbol{r} - \boldsymbol{r}') \tag{6.1.5}$$

where $g_l(\boldsymbol{r})$ is some coarse-graining function (a Gaussian, say) that leads to an average over a volume of linear dimension l in the integral (6.1.5) in order to remove the short-ranged part of the fluctuations. The correlation function for the potential due to this coarse-grained density fluctuation is expressed by

$$\overline{\langle \Delta V(\boldsymbol{r})\rangle_l \, \langle \Delta V(\boldsymbol{r}')\rangle_l} = \int \frac{d^3k}{(2\pi)^3} \, \bar{n}_\mathrm{A} \, v^2(\boldsymbol{k}) \, g_l^2(\boldsymbol{k}) \, \mathrm{e}^{\mathrm{i}\boldsymbol{k}(\boldsymbol{r}-\boldsymbol{r}')} \tag{6.1.6}$$

with the help of the Fourier transform $g_l(\boldsymbol{k})$ of the coarse-graining function. The divergence is not removed as the coarse-graining does not affect the long-ranged part of the Coulomb potential.

The divergence of the random potential can be understood easily: Let us consider subvolumes of our system of random acceptors, say cubes of linear dimension l. The average cube will contain $N_l = \bar{n}_\mathrm{A} \, l^3$ acceptors. This number will have a typical uncertainty $\Delta N_l = N_l^{1/2}$ due to the residual nonuniformity of a random distribution. The cubes will, therefore, be charged with a random charge of the order $e\Delta N_l$. If we consider these charges as concentrated at the centers of the cubes the resulting electrostatic potential on the surface of the cube will be of the order of

$$\Delta V = \frac{e^2}{4\pi\varepsilon_\mathrm{s}\varepsilon_0} \, \frac{\Delta N_l}{l/2} = \frac{e^2}{2\pi\varepsilon_\mathrm{s}\varepsilon_0} \, (\bar{n}_\mathrm{A} l)^{1/2} \tag{6.1.7}$$

which diverges as l is increased. Hence the coarse-grained potential of a system of fixed point charges distributed at random in space will always lead to such a divergency. The divergency is not removed by the positive random charges which we here treat as jellium. A treatment where charges of either polarity are considered as point charges distributed at random in space will also lead to a fluctuation of the actual net charges of subvolumes of the system.

The only mechanism by which the divergency can be removed (and by which they are removed in nature) involves mobile charges in addition to the fixed point charges. In our example there are

$$\bar{n}_\mathrm{e} = \bar{n}_\mathrm{D} - \bar{n}_\mathrm{A} \tag{6.1.8}$$

electrons left in the donor states that can hop between the donor states (or are mobile in the conduction band at elevated temperatures). It is clear that a fluctuation of length l can be completely screened if the density of these electrons exceeds the corresponding coarse grained density $\langle \Delta n_\mathrm{A} \rangle_l$. If we take a Gaussian as coarse-graining function we obtain

$$\overline{\left\langle \Delta n_\mathrm{A}(\boldsymbol{r}) \right\rangle_l^2} = \frac{1}{(4\pi)^{3/2}} \, \frac{\bar{n}_\mathrm{A}}{l^3} \tag{6.1.9}$$

and hence all fluctuations will be completely screened in case the length l exceeds l_max with

$$l_\mathrm{max} = \frac{1}{\sqrt{4\pi}} \, \frac{\bar{n}_\mathrm{A}^{1/3}}{\bar{n}_\mathrm{e}^{2/3}} \quad . \tag{6.1.10}$$

As a consequence we will have a random electrostatic potential for which the fluctuations with length l_{max} have the largest amplitude and for which the mean squared fluctuation δ^2 is given by

$$\delta^2 = 4\left(\sqrt{2}-1\right)\left(\frac{e^2}{4\pi\varepsilon_s\varepsilon_0}\right)^2 \frac{\bar{n}_A^{4/3}}{\bar{n}_e^{2/3}} \quad . \tag{6.1.11}$$

Note that the above derivation requires that the Fermi level is well below the conduction band and that the potential fluctuations are smaller than the donor binding energy. That this latter condition can be violated is easily seen from a numerical example. Suppose we have a semiconductor with $\varepsilon_s = 10$ and $\bar{n}_A = 10^{18}\,\mathrm{cm}^{-3}$ at a compensation level such that $\bar{n}_e = 5 \times 10^{15}\,\mathrm{cm}^{-3}$. From (6.1.10) we obtain $l_{max} = 960\,\text{Å}$ which is about 10 times the average distance between the impurities and the rms potential fluctuation is $\delta = 0.1\,\mathrm{eV}$, a value that clearly exceeds the binding energy of shallow donors in most semiconductors. Note that the typical volume for the larger fluctuations contains some 10^3 charged impurities. A typical fluctuation of the impurity density will be of the order of only 30 impurities in this volume which on the surface of the volume gives rise to a potential of the order of 0.05 eV, which is well compatible with a rms fluctuation of twice that value.

Although the above derivation is limited to semiconductors for which the compensation is nearly perfect (i.e., the acceptor density approximately equals the donor density) we have used it here because of its direct applicability to amorphous semiconductors. A more realistic treatment for the heavily doped compensated crystals is given in the book by Shklovskii and Efros [1984].

6.2 The Random Potential for Doped Amorphous Semiconductors

We have seen that a random potential requires the presence of a larger density of fixed charges and a small density of mobile charges. This condition is fulfilled in amorphous semiconductors because the density of states near the Fermi energy can accomodate most of the electrons or holes introduced by the electrically active dopants. The screening of the random potential will, therefore, not be achieved by mobile carriers.

Let us for the moment assume that we deal with a system that does not contain any charges in the absence of dopants. This means that the states below the Fermi energy E_{F0} for the undoped material are neutral if occupied while for the states above E_{F0} the opposite holds. If we now introduce donor states within the conduction band tail these become immediately ionized and the electrons will occupy states above E_{F0} which thereby become negatively charged. As a side-effect the Fermi energy is raised somewhat towards the conduction band and this shift of the Fermi energy is usually the motivation for doping.

We thus deal again with a system of point charges of both polarities: the positive donors usually are considered to be distributed randomly in space while for the negative charges this assumption probably does not hold. We concentrate for the moment on the positive charges of density n_+ and treat the negative charge as a jellium background that ensures overall neutrality of the sample. We thus obtain a random potential which modulates the energies of all states in a similar way. In particular this random potential will move deep states near midgap across the Fermi energy and thereby change their charge state. It is this redistribution of the charges due to the random fields that causes the screening. We may estimate the density of this screening charge by

$$n_{\mathrm{scr}} = g(E_{\mathrm{F}}) \cdot \delta \tag{6.2.1}$$

where δ is given by (6.1.11) if $\bar{n}_{\mathrm{A}}$ is replaced by the total density of positive charges, n_+, and if n_{e} is replaced by n_{scr}. We, therefore, can adapt Jäckle's formulae and obtain (omitting a factor of order unity)

$$l_{\max} = \left[\frac{\varepsilon_{\mathrm{s}}\varepsilon_0}{e^2\, g(E_{\mathrm{F}})}\right]^{1/2} \tag{6.2.2}$$

and

$$\delta = \left[\frac{e^2}{4\pi\varepsilon_{\mathrm{s}}\varepsilon_0}\right]^{3/4} \frac{n_+^{1/2}}{g(E_{\mathrm{F}})^{1/4}} \quad . \tag{6.2.3}$$

If we use the dielectric constant for silicon ($\varepsilon_{\mathrm{s}} = 12$) and the values $n_+ = 10^{18}\,\mathrm{cm}^{-3}$ and $g(E_{\mathrm{F}}) = 10^{16}\,\mathrm{cm}^{-3}\,\mathrm{eV}^{-1}$ for a numerical example we obtain the rather high values $l_{\max} = 2000\,\mathrm{Å}$ and $\delta = 0.2\,\mathrm{eV}$. Since $l_{\max}$ depends on the square root of the density of states near the Fermi energy the predominant potential fluctuations will be long-ranged even for rather high state densities.

In the above derivation we have assumed that the charges are distributed at random and that the potential is due to positive charges exclusively. In reality the value of δ must be twice the value given above if negative charges are also considered. The values will be enhanced drastically if the distribution of charges of one polarity is not perfectly random but clustered. The opposite holds of course if charges of opposite polarity are combined to form close pairs, i.e., dipoles. It is, therefore, not easy to estimate the value of δ for a real amorphous system.

A particular case exists if the Fermi energy is situated within a large dangling bond peak and if the effective correlation energy (see below in Sect. 7.1) is smaller than the width of this peak. In this case the density of charged centers will be approximately equal to the number of dangling bonds, n_{DB} and rather independent of the position of the Fermi energy. In this case we will have (note that here positive and negative charges must be situated at different centers)

$$\delta = \left[\frac{e^2}{4\pi\varepsilon_{\mathrm{s}}\varepsilon_0}\right]^{3/4} \frac{n_{\mathrm{DB}}^{1/2}}{g(E_{\mathrm{F}})^{1/4}} \tag{6.2.4}$$

which resembles the case of amorphous germanium (see Sect. 8.2).

If the effective correlation energy of the dominant defect level is negative the situation is the following: All the defect levels are charged, the Fermi energy is pinned between these charged states and the state density at the Fermi level can be quite small. This is supposed to be the case in the chalcogenide glasses for which we assume the potential fluctuations to be rather large.

Besides this electrostatic random potential there may be other origins for a long-ranged potential. Kramer et al. ([1983], see also King et al. [1983]) found a charge fluctuation connected with dangling bonds passivated by hydrogen. Knights and Lujan [1979] and Knights [1980] showed that under particular growth conditions the samples are quite inhomogeneous. Films grown under optimized conditions (i.e., those with a high photoconductivity) do not show this columnar structure if studied by electron microscopy but that is no guarantee for the inhomogeneities to be entirely absent [see Fritzsche, 1982]. Ley et al. [1982] found a broadening of the Si 2p core levels which they interpret as an effect of a random potential of 0.2 eV mean width.

When considering the effect of this random potential we note that there is a difference in length scales of at least 2 orders of magnitude between the short-range disorder and the long-ranged potential: The short-range disorder of the order of next nearest neighbour distances gives rise to the effects discussed at length in the previous two chapters. The long-ranged potential is superimposed on this system. If, as will be assumed throughout, the inelastic scattering length L_{i} of the carriers is small compared to l_{max}, the length scale of the random potential, a semiclassical treatment of the influence of this potential on the transport properties is possible.

6.3 Resistor Network Calculations

In order to evaluate the influence of the random potential on the transport parameters we shall go back to the simple Standard Transport Model. We shall assume that in the absence of the random potential the transport is above a well-defined mobility edge E_{C} with a constant prefactor σ_0 for the conductivity (and a constant heat of transport term, A, as well). The modifications discussed in the previous chapters will be postponed until later. The random potential, $V(\boldsymbol{r})$, will be assumed to be purely electrostatic modulating all energy levels in exactly the same way. We consider the case where the typical length l_{max} of the random potential exceeds L_{i}, the inelastic scattering length, by at least one order of magnitude. In thermal equilibrium the local carrier density will be constant in a cube of linear dimension exceeding a few times L_{i}. It will, therefore, be possible to define a local conductivity for this volume centered around X_i. Since

$$E_{\mathrm{C}}(X_i) - E_{\mathrm{F}} = \bar{E}_{\mathrm{C}} - E_{\mathrm{F}} + V(X_i) \tag{6.3.1}$$

we have a local conductivity which is given by

$$\sigma(X_i) = \bar{\sigma}_0 \exp(-\beta V(X_i)) \tag{6.3.2}$$

in terms of the values $\bar{E}_C$ and $\bar{\sigma}_0$ for the mobility edge and the conductivity, respectively, in the absence of the long-ranged potential. The local conductivity may be extended to a field $\sigma(\boldsymbol{r})$ by a smooth interpolation between the $\sigma(X_i)$ values.

In order to investigate how the rather broad distribution of local conductivities is combined to the conductivity of a macroscopic sample we divide the sample into small cubes. Within each cube the local conductivity is replaced by some mean value. The current through an array of conducting cubes can be replaced by a current through equivalent conductances, g_{ij}, connecting the center of a given cube with the centers of its neighbours.

$$g_{ij}^{-1} = \frac{1}{2g^0} \left[\exp(\beta V(X_i)) + \exp(\beta V(X_j)) \right] \tag{6.3.3}$$

where g^0 is the conductance of one cube in the absence of a random potential. The conductance of the resulting three-dimensional resistor network is evaluated solving Kirchhoff's equations [1845, 1847] for the potentials $V_i = V(X_i)$

$$\sum_{i \neq j} g_{ij}(V_i - V_j) = g_{i,\text{ext}} V_{\text{ext}} \tag{6.3.4}$$

where we have introduced conductances $g_{i,\text{ext}}$ that connect the cubes adjacent to the electrodes with the external potential, V_{ext}, and where the sums contain all the neighbours j of a given site i. From the solution V_i one may calculate the current through a plane parallel to the electrodes which finally gives the sample conductance, g.

The solution of Kirchhoff's equations requires the inversion of large matrices with random matrix elements that vary by many orders of magnitude. In our numerical work [Overhof and Beyer, 1981] we have been restricted to rather small model systems of $(10 \times 10 \times 10)$ cubes (most of the actual calculations were done on the smaller $(7 \times 7 \times 7$ samples)). In a first attempt we chose a cube length l comparable to l_{max} in which case the potentials of different cubes are essentially uncorrelated. It turned out, however, that results for a system consisting of cubes with truly random potentials are not at all representative of systems where the potentials are correlated (calculations for uncorrelated potentials have been performed as a check of the procedures, comparing the results to those of a percolation analysis). We, therefore, have chosen the parameters in such a way that L, the linear dimension of the sample, was $L = 2\, l_{\text{max}}$.

The construction of this potential started with a random distribution of point charges in the sample. The corresponding potential was averaged within each cube with periodic boundary conditions in every direction in order to cope with the boundary effects. The corresponding resistor network was constructed with periodic boundary conditions in the transverse directions. These conditions in combination automatically resulted in $L \simeq 2\, l_{\text{max}}$ without an explicit introduction of a screening length.

The potential thus constructed could be represented fairly well by a Gaussian distribution function of variance δ

$$p(V) = \frac{1}{\delta\sqrt{2\pi}} \exp\left(-\frac{V^2}{2\delta^2}\right) \quad . \tag{6.3.5}$$

Calculations using random numbers are usually plagued by numerical noise, i.e., by the fact that every set of random numbers gives a somewhat different result. An obvious way out of this problem is a configurational average, i.e., a repetition of the calculation with several different sets of random numbers and a proper average of the results thus obtained. Care must be taken to ensure that the configurational average does not introduce artificial new structures into the results but serves to improve the reliability of the resulting numbers only.

The proper average presents a real problem in our case because we have no obvious recipe how the average should be performed. An arithmetic average of the sample conductances is equivalent to a sheet of conductances in parallel while an arithmetic average of the resistances would correspond to a chain of resistances in series. In the former case (particularly in calculations for lower temperatures) a sample with a highly conducting channel will dominate the average while in the latter case the result will be determined essentially by samples with highly resistive barriers. As both averages do not resemble the system of interconnected networks we use the geometric average for the conductance calculations.

The resulting conductance shows surprisingly small temperature dependence. Figure 6.1 shows an Arrhenius plot of the conductance for various values of δ. For lower temperatures the conductance can be approximated (apart from a trivial geometry factor) by

$$\ln(g/g^0) = -0.2 + 0.35\,\delta/kT \quad . \tag{6.3.6}$$

We see that the predominant transport path is mainly through "valleys" of lower potential and determined by "cols" where the potential is still slightly negative.

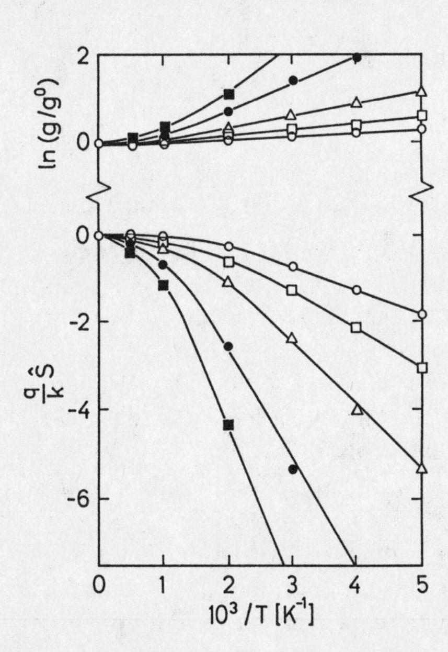

Fig. 6.1. Calculated average network conductance (*upper part*) and thermoelectric power (*lower part*) as a function of inverse temperature for $10 \times 10 \times 10$ resistor networks with long-ranged potentials due to point charges of different densities

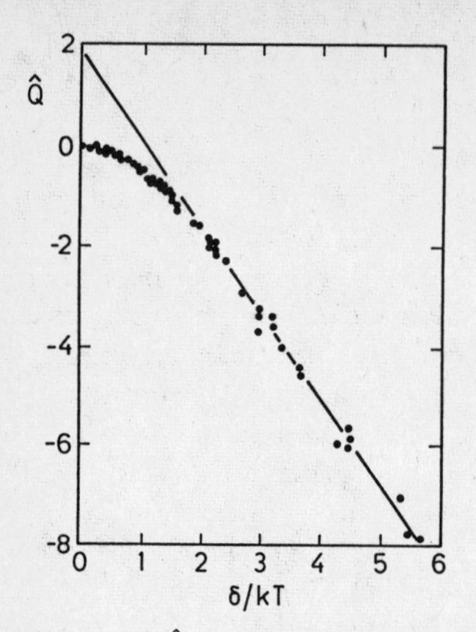

Fig. 6.2. $\hat{Q}$ function derived from the model of Fig. 6.1 as a function of the inverse temperature normalized by the variance of the random potential

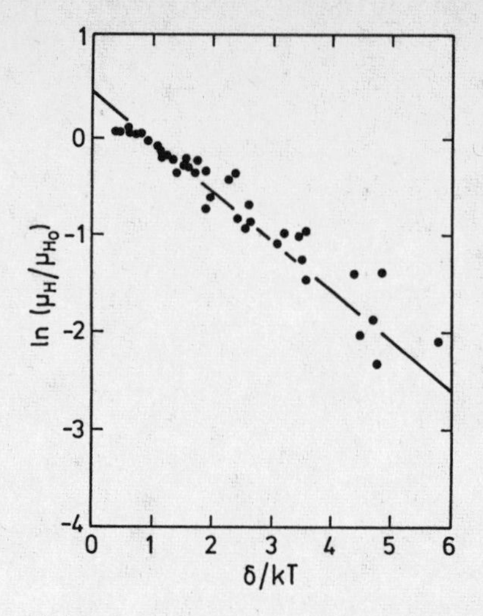

Fig. 6.3. Averaged Hall mobility calculated for the $10 \times 10 \times 10$ resistor model as a function of the inverse temperature normalized by the variance of the random potential

For elevated temperatures the conductance deviates from (6.3.6) and becomes essentially constant.

We calculate the Peltier heat in the same way and obtain the thermoelectric power using Onsager's symmetry. The result, also shown in Fig. 6.1, can be represented by (for $\delta > 3kT$)

$$\frac{q}{k}\, S(T) = \frac{E_{\mathrm{C}} - E_{\mathrm{F}}}{kT} + A + 2 - 2.1\,\frac{\delta}{kT} \tag{6.3.7}$$

which can be combined with (6.3.6) to yield

$$Q(T) = \ln(\sigma_0\, \Omega\, \mathrm{cm}) + A + 1.8 - 1.75\,\frac{\delta}{kT} \quad . \tag{6.3.8}$$

The Q function, therefore, is "activated" with $E_Q^* \simeq 2\delta$. This is exemplified in Fig. 6.2 where $\hat{Q} = Q - \ln(\sigma_0\, \Omega\, \mathrm{cm}) - A$ is plotted versus δ/kT. The figure shows that our numerical model calculation leads to consistent results.

One particular point should be emphasized here: The Arrhenius plot of most properties shows for lower temperatures a linear behaviour. At elevated temperatures the effect of the long-ranged potential becomes negligible. Consequently the curves bend over and the properties become temperature independent. This transition leads to a nonzero intercept of the extrapolated low temperature curves – the extrapolated value is different from the value at $1/T = 0$. This explains why an additional term 1.8 occurs in (6.3.8).

It is possible to extend the resistor network model to treat the Hall mobility [Overhof, 1981]. If the Hall mobility has the constant value μ_{H0} in the absence of the long-ranged potential one obtains in the presence of this random potential

$$\ln(\mu_{\rm H}/\mu_{\rm H0}) = 0.25 - 0.6\,\delta/kT \tag{6.3.9}$$

as is plotted in Fig. 6.3. This result would perfectly compare with experimental data [Overhof, 1981], in particular the correlation of the apparent activation energies of Q and the Hall mobility

$$E^*_{\rm H} \cong \frac{1}{3}\,E^*_Q \quad . \tag{6.3.10}$$

Unfortunately the sign anomaly of the Hall effect cannot be explained at all by a random potential and, therefore, the agreement does not bring us nearer to an understanding.

6.4 Random Walk Computations

The resistor network computations are limited to rather small samples because of the requirement of a matrix inversion. With the availability of bigger computer facilities these computations can be extended to larger samples. The progress, however, is rather small: If the linear dimension of a sample is increased by a factor of two the rank of the matrix to be inverted is increased by a factor of 8 and the computational effort required is raised by a factor of 512.

In order to overcome these severe sample size limitations we, therefore, concentrate on random walk calculations. These computations have been performed on samples consisting of N × N × M cubes (with N of order 20 – 50 and M in the 200 – 1000 range). For these rather largish samples the random potential could not be constructed in the time-consuming way described in the previous section. Instead we have constructed a random potential by the following procedure: We start assigning random potential values to each cube. As a next step the potential value for each cube is determined from the initial potential values averaging over the 27 neighbours of a cube (with periodic boundary conditions). Iteration of this procedure leads to a potential that has a finite correlation length $l_{\rm max}$ which is proportional to the square root of the number of iteration steps and a distribution function $p(V)$ given by (6.3.5). The potential is then normalized to $\delta = 0.15\,{\rm eV}$ and an external potential in the M-direction is added.

Carriers are generated in the interior of the sample and their movement through the sample is monitored under the combined action of the random potential and the external potential. The movement of the carrier is calculated assuming that the transition probability for a transit from cube i to cube j is given by

$$p_{ij} = \begin{cases} \exp(-\beta\,\Delta V_{ij}) & \Delta V_{ij} \geq 0 \\ 1 & \Delta V_{ij} < 0 \end{cases} \tag{6.4.1}$$

if ΔV_{ij} is the total potential difference between adjacent cubes. At each position of the carrier the next step is determined by a random number weighted by the transition probabilities to the different neighbours while the inverse of the sum of the transition probabilities is taken as the characteristic time for the transit.

The procedure is then iterated until the carrier reaches one of the electrodes. A particular difficulty arises if we treat lower temperatures. If the carrier comes to a local minimum of the potential the back and forth movement between the minimum and the adjacent sites can consume enormous quantities of computer time (even if the actual time spent by the carrier is not exceptionally large). This complication could be eliminated by an algorithm in which the average total time spent in the minimum was calculated first and a direct return to this minimum was excluded.

6.4.1 Conductivity and Thermoelectric Power

For the calculation of the transport properties of carriers in quasiequilibrium we have to assume that the carrier concentration in some cube is proportional to

$$n_{\mathrm{C}}(X_i) = N_{\mathrm{C}} \exp(-\beta V(X_i)) \qquad (6.4.2)$$

if $V(X_i)$ is the random potential. This causes problems with the configurational average if calculations are performed for lower temperatures. For these calculations, therefore, the Gaussian distribution function for the potential was cut off for $|V(X_i)| > 3\delta$.

For the lower applied fields the configurationally averaged current was Ohmic and the temperature dependence of this current was similar to that found in the resistor network calculations. In a similar way the Peltier coefficient has been calculated. The resulting $\hat{Q}$ data shown in Fig. 6.4 are essentially the same as those shown in Fig. 6.3 except that the resulting slopes and intercepts are somewhat different. We find from these data that

$$Q = \ln(\sigma_0\, \Omega\, \mathrm{cm}) + A + 3.5 - 1.1\,\frac{\delta}{kT} \qquad (6.4.3)$$

if $\delta > 3\,kT$ which is in reasonable agreement with the results of the resistor network calculations.

6.4.2 Superlinear $I - V$ Characteristics

The random walk model allows for the simple extension for the applied external fields to be increased far beyond the Ohmic region. Once this field is no longer small compared to the internal field we expect a deviation from an Ohmic current to a superlinear field dependence. At the other extreme, if the applied field is stronger than the internal random potential we expect again a transition to an Ohmic regime. Results are shown in Fig. 6.5 where the data have been calculated for a sample with $l_{\max} = 1000$ Å, $\delta = 0.14$ eV, and a sample thickness of 1 μm. From an inspection of these data we observe that the steepest parts of the $I - V$ characteristics can be represented by a power law

$$I \sim V^{l_{\mathrm{SLC}}} \qquad (6.4.4)$$

with an exponent (we use the abbreviation SLC in order to discriminate our superlinear current from a SCLC current)

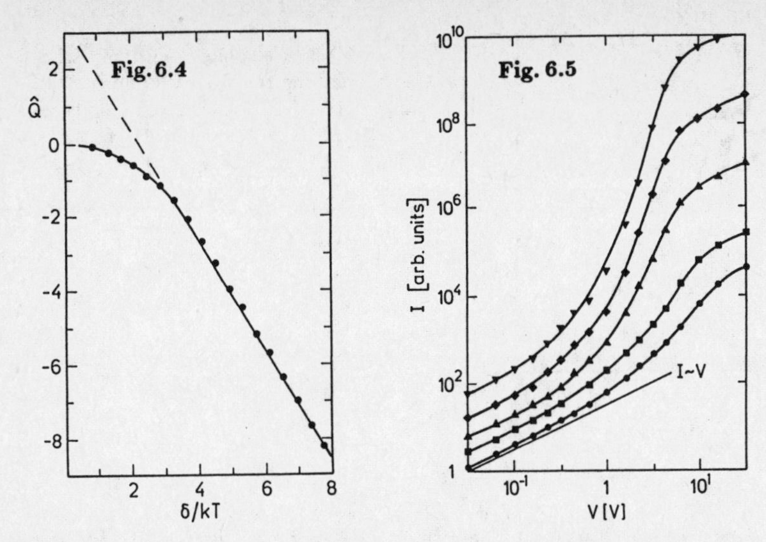

Fig. 6.4. $\hat{Q}$ function calculated for the $50 \times 50 \times 200$ random walk model as a function of the inverse temperature normalized by the variance of the random potential

Fig. 6.5. $I - V$ characteristic calculated for the $50 \times 50 \times 200$ random walk model with $\delta = 0.15\,\mathrm{eV}$, $l_{\mathrm{max}} = 1000\,\mathrm{Å}$, and $l = 1\,\mu\mathrm{m}$. Note that there is no space charge effect included in the model. Temperatures are 200 K (▼), 250 K (♦), 333 K (▲), 500 K (■), and 666 K (●), respectively

$$l_{\mathrm{SLC}} = 1 + T_{\mathrm{SLC}}/T \tag{6.4.5}$$

with $T_{\mathrm{SLC}} = 1100\,\mathrm{K}$ as is frequently observed in experiments (which, however, usually are interpreted in terms of SCLC). The shape of the $I - V$ characteristic is also practically indistinguishable from those calculated for the SCLC except for the highest current levels (see [Overhof, 1985] for details). A variation of the sample thickness, however, does not lead to the standard scaling law which must be observed for every SCLC. Instead we find from our data a scaling law that approximately reads

$$I/l = f(V/l^{1.4}) \tag{6.4.6}$$

in contrast to (3.4.7). As mentioned above the state of the art of present sample preparation methods does not lead to samples of sufficient reproducibility to distinguish between both sets of power laws.

6.4.3 Time-of-Flight Data

The time dependence of the current due to carriers moving in the random field is of particular interest. We treat the case where the probability to create a carrier at some point $\boldsymbol{r}$ in space does not depend on the value of the random potential at that point as is the case for, e.g., optical excitation of carriers. This con-

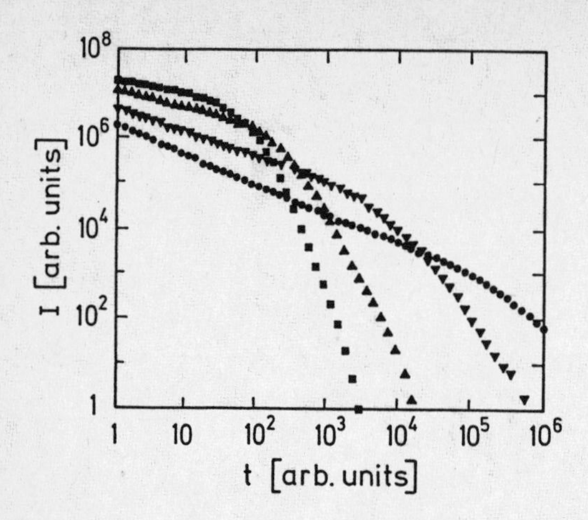

Fig. 6.6. Calculated current transients for the random walk model with $\delta = 0.15\,\mathrm{eV}$ for different temperatures ($T = 125\,\mathrm{K}$ (●), $T = 175\,\mathrm{K}$ (▼), $T = 250\,\mathrm{K}$ (▲), $T = 300\,\mathrm{K}$ (■))

siderably simplifies the configurational averages. Representative results for the configurationally averaged current transients are shown in Fig. 6.6. Note that the anomalous dispersion observed is not an artefact of the configurational average: Results obtained for a single sample show qualitatively the same transients, the configurational averages simply reduce the statistical noise. We stress this point because dispersive transients are due to a wide variation of characteristic times in the system, and it is, therefore, necessary to check that the distribution of characteristic times is not generated by the configurational average. Such an error was, e.g., present in the earlier Monte–Carlo simulations of dispersive transients for R-hopping systems (see, e.g., Adler et al. [1982]). We find transients that show a marked break if $\log(I)$ is plotted as a function of $\log(t)$. For times smaller than this break the transients very closely follow the algebraic law

$$I(t) \sim t^{-(1-\alpha_1)} \tag{6.4.7}$$

with $\alpha_1 \sim T$ while for times larger than the break the curves only approximately follow

$$I(t) \sim t^{-(1+\alpha_2)} \tag{6.4.8}$$

with a rather broad transition range following the break. In contrast to our earlier calculations [Overhof, 1983] the results shown in Fig. 6.6 have been calculated with the high temperature transition probabilities (6.4.1). If we define the transit time by the intersection of the straight lines of a power law fit (see, e.g., Seynhaeve et al. [1988] for a discussion of this procedure) we find an activated drift mobility with an activation energy close to the potential width parameter δ. Note that for small applied fields F the transit times according to our calculation are inversely proportional to this field. This seems to be in conflict to most TOF experiments for undoped a-Si:H samples in the anomalously dispersive regime. Here the observed transit times are proportional to $F^{-1/\alpha}$ in general agreement with the predictions of the TROK model.

We just mention that a Fourier transform of $I(t)$ should give the ac conductivity if the current transients are calculated using the equilibrium carrier distribution function. Unfortunately, this has not yet been done but it might be of considerable interest if experimental TOF and ac conductivity data on comparable samples would be available.

7. Temperature Dependent Reference Energies

In the previous sections we have always assumed that the electronic states, the Fermi energy, etc., do not depend on temperature. We have seen, however, in Chap. 4 that the energy of the centroid of the carriers in thermal equilibrium somewhat changes with rising temperatures. We also know from optical experiments that the optical band gap decreases with rising temperature (the "red shift of the band gap", c.f. Sect. 2.1). These effects shall now be taken into account and their influence on the observed transport properties shall be discussed. To begin with, we shall show in the first section of this chapter that there is always a significant temperature dependence of the Fermi energy, called the statistical shift of the Fermi energy. It turns out that this effect alone explains the observed Meyer–Neldel rule even if quite general density of states distributions are considered. We shall then discuss in the second section the controversial matter how the red shift of the band gap caused by the electron-phonon coupling influences the transport properties. In both sections we can discuss the essential points using our Standard Transport Model discussed in Sect. 2.6.

7.1 The Statistical Shift of the Fermi Energy

In the statistical mechanics for a fermion system the Fermi energy E_F enters as a Lagrangian parameter that determines the mean particle density. In a solid the position of this energy must be such that the total density of positive charges due to the ions that form the solid is balanced by the density of valence and conduction electrons (including defect ions and electrons in defect states). Due to the laws of electrostatics the overall neutrality of a macroscopic body must be observed with extreme perfection: a net charge of 10^{10} excess electrons on a body of 1 cm length leads to an electrostatic potential of this body that amounts to 10^3 Volts.

For a gas of noninteracting electrons the number density of negative charges is given by an integral over the density of states times the Fermi distribution function

$$n_{\text{tot}} = \int_{-\infty}^{\infty} 2 \cdot g(E) \cdot f_F(E, E_F, T)\, dE \tag{7.1.1}$$

where the Fermi–Dirac distribution function already discussed in Sect. 2.6 is given here with all its arguments. A factor of 2 enters because of the spin degeneracy of the states. This density is exactly equal to the number density of positive nuclear charges. If we treat deeply localized states in the pseudogap we have to consider correlation effects. Due to the electrostatic Coulomb repulsion the energy of the localized states will depend on the occupancy. The simplest and most widely used model for correlation in localized states has been given by Hubbard [1964]: Interaction of electrons (with opposite spin σ) localized at the same "site" is described by a Hamiltonian that consists of single particle energies ε_i and of correlation energies U_i

$$H = \sum_{i,\sigma} n_{i\sigma}\varepsilon_i + \sum_i U_i n_{i\downarrow} n_{i\uparrow} \qquad (7.1.2)$$

which enter for the doubly occupied states only.

A more detailed treatment must further take into account the electron-phonon coupling which may lead to a significant distortion of the lattice around a defect depending on the defect's occupancy. The distortion due to the first electron occupying a given site can be large and attractive for a second electron of opposite spin [Anderson, 1975, 1976].

We shall demonstrate the effect following a simplified model due to Liciardello et al. [1981, see also Liciardello, 1981] which considers local phonon modes exclusively. These modes with frequencies ω_i are localized at some site i and the Hamiltonian for these phonons can be expressed in terms of operators $b_i^\dagger$

$$H_\mathrm{p} = \sum_i \hbar\omega b_i^\dagger b_i \qquad (7.1.3)$$

with a coupling to the electron system which is given by

$$H_\mathrm{ep} = \sum_{i,\sigma} \hbar\omega A_i (b_i^\dagger + b_i) n_{i\sigma} \quad . \qquad (7.1.4)$$

The Hamiltonian for the coupled system

$$H = H_\mathrm{e} + H_\mathrm{p} + H_\mathrm{ep} \qquad (7.1.5)$$

can be diagonalized after a canonical transformation leading to displaced harmonic oscillator coordinates

$$d_i = b_i + A_i \sum_\sigma n_{i\sigma} : \qquad (7.1.6)$$

$$H = \sum_{i,\sigma} (\varepsilon_i - E_{\mathrm{b}i}) n_{i\sigma} + \sum_i (U_i - 2E_{\mathrm{b}i}) n_{i\uparrow} n_{i\downarrow} + \sum_i \hbar\omega\, d_i^\dagger d_i \quad . \qquad (7.1.7)$$

The Hamiltonian describes quasiparticles that are decoupled from the displaced harmonic oscillators. The quasiparticles are bound to the sites by an extra binding energy

$$E_{\mathrm{b}i} = \hbar\omega A_i^2 \quad . \qquad (7.1.8)$$

Two quasiparticles on a single site interact with the effective correlation energy

$$U_{\mathrm{eff},i} = U_i - 2E_{\mathrm{b}i} \tag{7.1.9}$$

which may be negative, i.e., attractive. Watkins and Troxel [1980] present evidence that the vacancy in crystalline silicon is a defect with negative effective correlation energy.

For chalcogenide glasses Street and Mott [1975] and Mott et al. [1975] have shown that the predominant defects have a negative correlation energy which in this particular case, however, is due to chemical bonding. For tetrahedrally bonded amorphous semiconductors the presence of paramagnetic states (detected by electron spin resonance) proves that at least for these latter states the effective correlation energy must be positive (see, however, Adler [1981, 1987]).

7.1.1 The Statistical Shift for Some General Density of States Models

Independent of the sign of the effective correlation energy we obtain the occupation probabilities that a site is unoccupied, singly occupied, or doubly occupied from the partition function Z.

$$f_1(E, E_{\mathrm{F}}, T) = \frac{2}{Z} \exp[-\beta(E - E_{\mathrm{F}})] \tag{7.1.10}$$

gives the probability for a site of energy E to be singly occupied where the factor of 2 is again due to the two possbile spin orientations and

$$Z = 1 + 2 \cdot \exp[-\beta(E - E_{\mathrm{F}})] + \exp[-\beta(2E - 2E_{\mathrm{F}} + U_{\mathrm{eff}})] \quad . \tag{7.1.11}$$

The probability that two electrons occupy the same site is given by

$$f_2(E, E_{\mathrm{F}}, T) = \frac{1}{Z} \exp[-\beta(2E - 2E_{\mathrm{F}} + U_{\mathrm{eff}})] \tag{7.1.12}$$

and the neutrality condition reads

$$n_{\mathrm{tot}} = \int_{-\infty}^{\infty} g(E)\,[f_1(E, E_{\mathrm{F}}, T) + 2 \cdot f_2(E, E_{\mathrm{F}}, T)]\,dE \tag{7.1.13}$$

which reduces to (7.1.1) if U_{eff} is set equal to zero.

The distribution functions are illustrated in Figs. 7.1 and 7.2 for a positive value of the correlation energy U_{eff}. Note that the functions are essentially step-like with a transition range of width kT. There is a long tail of f_1 extending to higher positive energies which can be approximated by a Boltzmann distribution

$$f_1(E, E_{\mathrm{F}}, T) \simeq \exp[-\beta(E - E_{\mathrm{F}})] \quad . \tag{7.1.14}$$

This tail is responsible for the activated density of carriers in the conduction band (the same holds for $[1 - f_2(E, E_{\mathrm{F}}, T)]$ which gives the probability for a hole in the valence band).

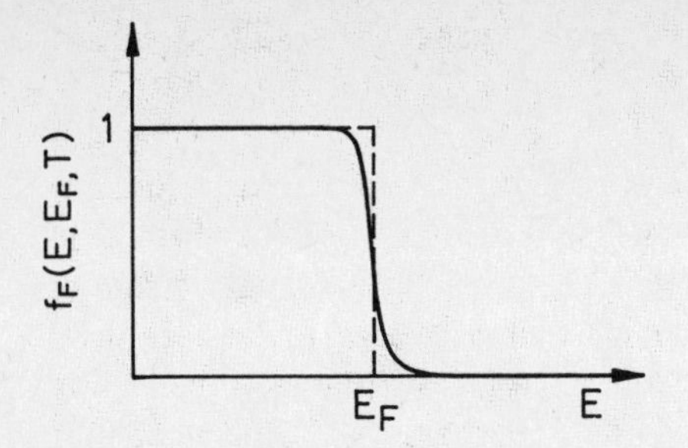

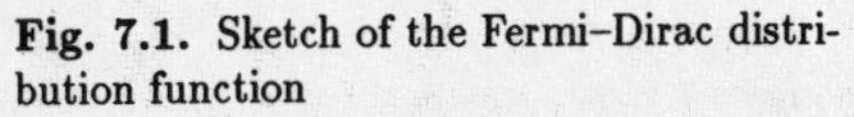
Fig. 7.1. Sketch of the Fermi–Dirac distribution function

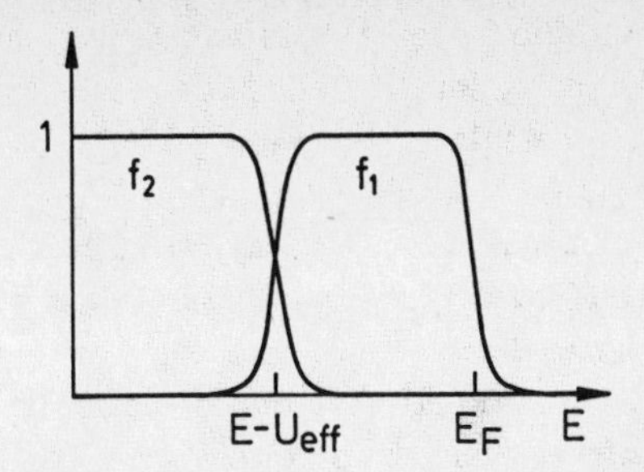

Fig. 7.2. Sketch of the distribution functions f_1 and f_2 for states with a positive effective correlation energy, U_{eff}

For crystalline semiconductors it is well known that the neutrality condition requires that the Fermi energy moves with temperature towards midgap ("statistical shift of the Fermi energy", Busch and Labhart [1946], see also Madelung [1957]). This is exemplified in Fig. 7.3 where the density of states distribution near the gap of a doped crystalline semiconductor is sketched. At very low temperatures all the donor states will be occupied and the Fermi energy will be just below the donor level E_{D}. With rising temperature the donors will be gradually ionized. Once the ionization of the donors is complete the Fermi energy has to shift away from the conduction band in order to ensure that the density of carriers in the conduction band does not exceed the density of ionized donors. This shift of the Fermi energy continues in the "exhaustion regime" until (in the "intrinsic regime") the Fermi level is close to midgap. The neutrality of the crystal is then achieved by the simultaneous generation of additional electrons in the conduction band and holes in the valence band.

For amorphous semiconductors we expect a significant shift of the Fermi energy (Spear and LeComber [1976], Beyer et al. [1977]) with temperature if the density of states is highly asymmetric with respect to the Fermi energy. As a simple example we show in Fig. 7.4 a density of states distribution consisting of a conduction band with a steep tail and a constant density g_0 of deep states in the pseudogap (the shape of the valence band is of no practical importance if

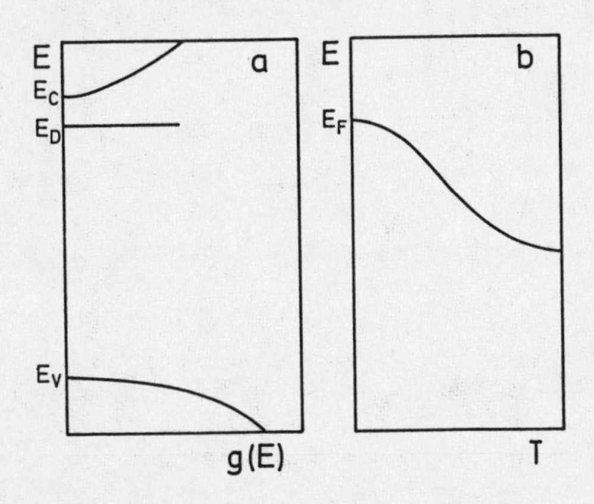

Fig. 7.3. (**a**) Schematic density of states distribution for an n-type crystal (**b**) Position of the Fermi energy as a function of temperature

we treat the statistical shift in the upper half of the pseudogap). The resulting $E_F(T)$ curves for different positions of E_F at $T = 0$ are shown in Fig. 7.5: the full line was calculated for $g_0 = 10^{15}\,\mathrm{cm^{-3}\,eV^{-1}}$ while the dashed-dotted line was obtained with $g_0 = 10^{17}\,\mathrm{cm^{-3}\,eV^{-1}}$. These curves are rather typical for density of states distributions characterized by a steep tail and a small density of states in the pseudogap. We note in passing that curves like those shown in Fig. 7.5 cannot be compared directly to results of a sequence of differently doped samples unless it can be assumed that doping leads to a modification of the electron density exclusively. This assumption may be valid if, e.g., doping only introduces donor states in the upper part of the tail. It will certainly not be fulfilled if doping also changes the density of dangling bond states as must be assumed for a-Si : H (see Sect. 8.1.1).

In the case that the density of states distribution can be expanded into a power series around the zero temperature Fermi energy it is possible to write down analytical formulae for the statistical shift. If we expand

$$g(E) = g(E_{F0}) + g'(E_{F0}) \cdot (E - E_{F0}) \tag{7.1.15}$$

we can calculate the electron density for a fixed Fermi level position at E_{F0}. This leads us to (again we have a factor of 2 due to spin degeneracy)

$$n = n_0 + g'(E_{F0})\,\frac{\pi^2}{3}\,(kT)^2 \tag{7.1.16}$$

where the temperature dependence of the electron density reflects the fact that the Fermi level was considered fixed. This temperature dependence of the electron

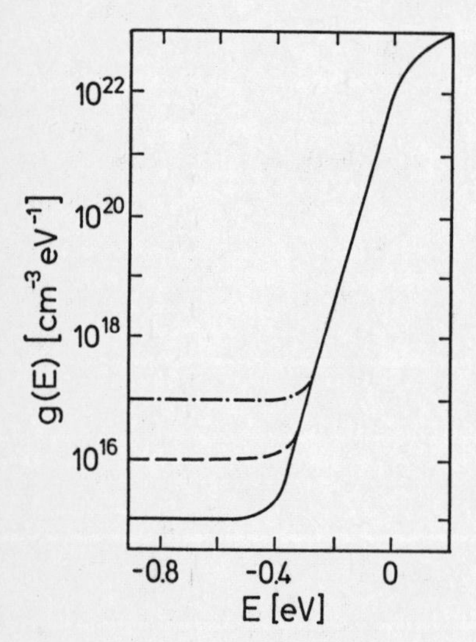

Fig. 7.4. Simple model density of states for hydrogenated amorphous semiconductors with a small density of midgap states and a steep conduction band tail

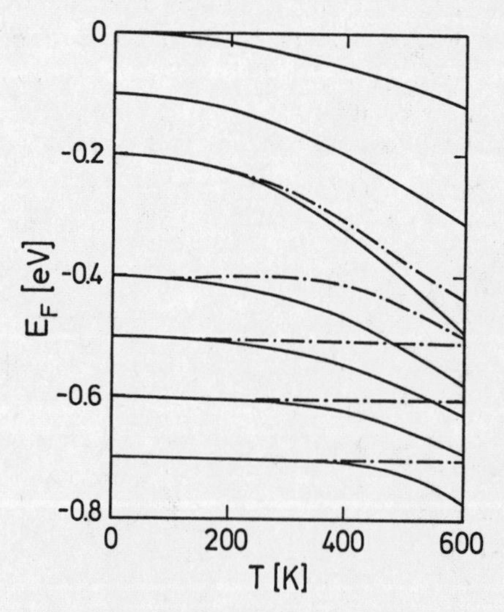

Fig. 7.5. Statistical shift of the Fermi energy for the two extreme density of states distributions of Fig. 7.4. The location of $E_F(T = 0)$ is taken as a free parameter

density must be balanced by a temperature dependent displacement of the Fermi level by

$$\delta E_{\mathrm{F}} = -\frac{g'(E_{\mathrm{F0}})}{g(E_{\mathrm{F0}})}\,\frac{\pi^2}{6}\,(kT)^2 \tag{7.1.17}$$

which may be differentiated to yield

$$\frac{dE_{\mathrm{F}}}{dT} = -\frac{\pi^2}{3}\,k^2 T\,\frac{g'(E_{\mathrm{F0}})}{g_0(E_{\mathrm{F0}})} \quad . \tag{7.1.18}$$

This formula which is very useful for the determination of the Fermi level shift in metals just accounts for the redistribution of the electrons within kT around the Fermi level due to the broadening of the Fermi–Dirac distribution function. As is appropriate for metals it disregards the extreme wings of the distribution function which, however, are of utmost importance in our case. These wings extend into an energy range where the density of states distribution rises by several orders of magnitude. From the expansion (7.1.15) we see that the validity of the metallic formula is limited to the range of parameters which ensure that $\delta E_{\mathrm{F}} < kT$. It is, therefore, usually misleading to conclude from a large statistical shift that the density of states at the Fermi energy must be strongly energy dependent. There are numerous examples where at elevated temperatures the Fermi level shifts into the opposite direction than that anticipated from (7.1.18).

7.1.2 Implications on Conductivity and Thermoelectric Power

We may approximate the calculated $E_{\mathrm{F}}(T)$ curves by a linear relation in a limited temperature range (usually from about 300 K to 400 K) and obtain (see Fig. 7.6)

$$E_{\mathrm{F}}(T) \cong E^*_{\mathrm{F}}(0) + \gamma^*_{\mathrm{F}} \cdot T \tag{7.1.19}$$

where the exact values for E^*_{F} and γ^*_{F} depend somewhat on the temperature range used for the linear approximation. If we insert (7.1.19) into the formulae of Sect. 2.6 we obtain

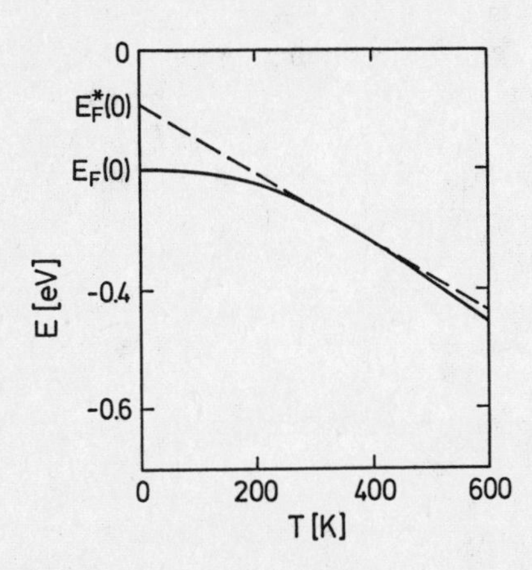

Fig. 7.6. Sketch of a typical statistical shift of E_{F} and its linear approximation

$$\sigma(T) = \sigma_0^* \exp(-E_\sigma^*/kT) \tag{7.1.20a}$$

with

$$\sigma_0^* = \sigma_0 \exp(\gamma_F^*/k) \tag{7.1.20b}$$

and

$$E_\sigma^* = E_C - E_F^*(0) \tag{7.1.20c}$$

which shows that the statistical shift of the Fermi energy leads to a modification of the apparent activation energies as well as of the observed prefactors as already discussed by Fritzsche [1971]; see also [Jones et al., 1977; Beyer et al., 1977; Mott and Davis, 1971, 1979]. Note that according to (7.1.19) and (7.1.20c) the difference ΔE_σ^* between the observed activation energy E_σ^* and the "true activation energy" $E_C - E_F(T)$ is given by

$$\Delta E_\sigma^* = E_\sigma^* - [E_C - E_F(T)] = \gamma_F^* \cdot T \tag{7.1.20d}$$

which usually amounts to several kT.

A similar analysis for the thermoelectric power leads to

$$\frac{q}{k} S(T) = E_\sigma^*/kT + A^* \tag{7.1.20e}$$

(note that according to the Standard Transport Model the activation energies of the thermoelectric power and of the conductivity are identical) with

$$A^* = A - \gamma_F^*/k \tag{7.1.20f}$$

i.e. slopes and intercepts of conductivity and thermoelectric power are affected in exactly the same way and hence in $Q(T)$ the effect is totally invisible. In fact the complete insensitivity of $Q(T)$ to the statistical shift of the Fermi energy (which is generally not exactly known) lead Beyer and Overhof [1979] to introduce this function for the analysis of the transport mechanism.

If we start with the $E_F(T)$ curves of Fig. 7.5 and derive from the approximation (7.1.19) the apparent activation energies and the averaged shift parameter we obtain a "Meyer–Neldel rule" as shown in Fig. 7.7. Depending on the details of the density of states distribution function we calculate somewhat different curves which, however, are similar in the following points: There is a range of E_σ^* values for which γ_F^* varies approximately linearly with the apparent activation energy. If we consider the microscopic prefactor of the conductivity to be a constant we derive the following Meyer–Neldel rule from the Fig. 7.7

$$\ln(\sigma_0^* \,\Omega\,\mathrm{cm}) = B_{\mathrm{MNR}} + E_\sigma^*/E_{\mathrm{MNR}} \tag{7.1.21}$$

as in (3.1.1) with the numerical value for B_{MNR}

$$B_{\mathrm{MNR}} = \ln(\sigma_0 \,\Omega\,\mathrm{cm}) - 12 \pm 2 \tag{7.1.22}$$

and $E_{\mathrm{MNR}} = 0.043\,\mathrm{eV}$.

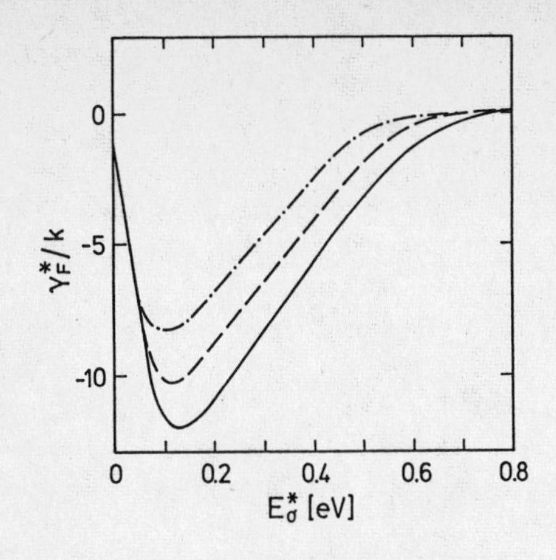

Fig. 7.7. Mean Fermi level shift in the $250\,\mathrm{K} < T < 350\,\mathrm{K}$ interval as a function of the apparent activation energy E_σ^* calculated for the model densities of states shown in Fig. 7.4

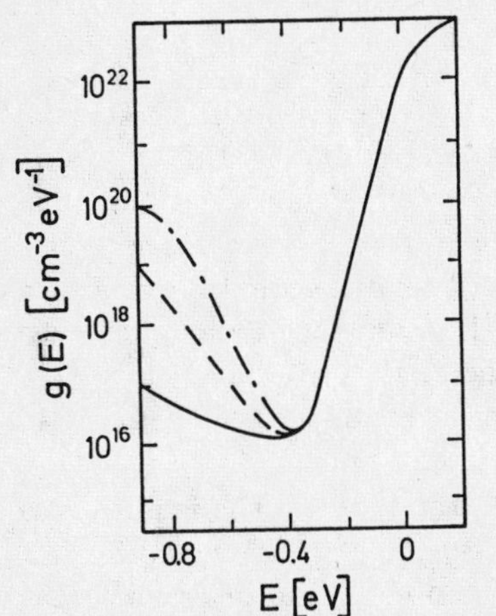

Fig. 7.8. Alternative density of states distribution with a minimum in the density of states below the conduction band tail

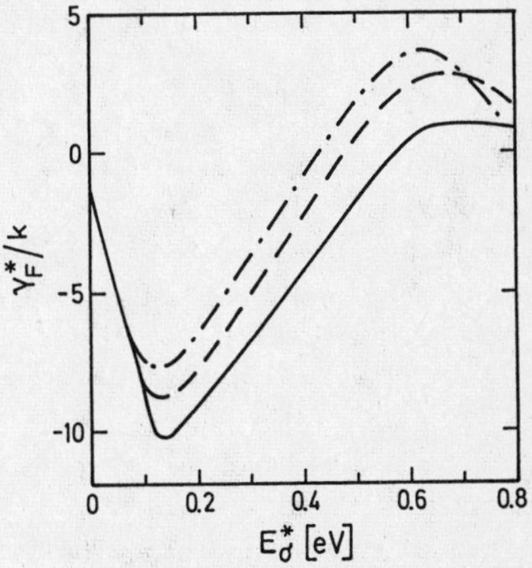

Fig. 7.9. Mean statistical shift for the Fermi level calculated from the density of states model of Fig. 7.8 in the $250\,\mathrm{K} < T < 350\,\mathrm{K}$ temperature range

This explanation for the Meyer–Neldel rule was proposed first by Overhof and Beyer [1981]; see also [Overhof and Beyer, 1983; Drüsedau and Bindemann, 1986; Drüsedau et al. 1987]. The interpretation, however, was doubted by Irsigler et al. [1983] because a specific density of states distribution was used to calculate the statistical shift which seemed to contradict the finding of the ubiquitous presence of the MNR.

We should, therefore, like to stress the fact that a MNR as shown in Fig. 7.7 is not limited to a particular density of states distribution but always results if the ratio of conduction band states to midgap states is large. Fig. 7.8 shows a density of states model where the energy independent density of deep midgap states of the previous model has been replaced by an exponential distribution of deep states forming a minimum in the density of states just at the onset of the tails. We see from the resulting MNR in Fig. 7.9 that in fact the details of the density of states distribution do not influence the general shape of the resulting Fermi level shift.

Our interpretation of the MNR leads to a consequence that can be checked experimentally: The observed values for B_{MNR} and for E_{MNR} depend on the temperature range for which the Arrhenius plot of the conductivity is evaluated. If this range is changed to lower temperatures the slope E_{MNR} will be decreased and the intercept B_{MNR} will ultimately approach $\ln(\sigma_0\,\Omega\,\mathrm{cm})$.

Strangely enough quite similar Meyer–Neldel rules are observed for the dispersive diffusion of impurities in crystals [Narasimhan and Arora, 1985] and also for the dispersive transport of excess carriers in amorphous semiconductors. Jackson [1988] has shown that in the latter case the MNR can be explained by the wide distribution of event times, release from traps in this case. A similar mechanism could also be valid in the former case.

7.1.3 The Meyer–Neldel Rule of the SCLC

We have just shown that we obtain a Meyer–Neldel rule if we assume that for a given density of states distribution the Fermi level position (at $T = 0$, say) can be altered. This is exactly the case in SCLC experiments if the simplifying assumption is made that the induced charge can be considered homogeneously distributed in the sample. We see from (3.4.3) that in the limiting case of a homogeneous distribution of the space charge in the sample there is no difference between the statistical shift of the quasi-Fermi energy for a fixed applied voltage (i.e., for a fixed total density of electrons) and the shift of the true Fermi level for a system with the same electron density in equilibrium.

In the evaluation of some experiments the Fermi level shift is ignored and the change of the apparent activation energy E_{a}^* in (3.4.8) is taken as the shift of the quasi-Fermi energy $E_{\mathrm{F}}^{\mathrm{q}}$. The result of this procedure can be estimated as follows: with $E_{\mathrm{MNR}} = 0.05\,\mathrm{eV}$ and $T \simeq 300\,\mathrm{K}$ we find that the change of E_{a}^* is about twice the corresponding change of $E_{\mathrm{F}}^{\mathrm{q}}$ and hence the inferred density of states will be too small by a factor of two. For smaller values of E_{MNR} or measurements at elevated temperatures the effect will be even larger.

While an error in the density of states by a factor of two might be considered tolerable the incorrect determination of the Fermi energy by the apparent activation energy is in most cases most undesirable: A change of the apparent activation energy is in several instances taken as a "direct experimental determination" of the change of the Fermi energy itself. In most cases, however, the true change of the Fermi energy amounts to at most half that value. Parameters like positions of the Fermi energy with respect to the mobility edge cannot be determined directly from the experiment. Instead, one has to start with a theory that matches the experiments to a high degree of perfection. The parameters which enter this theory can then be identified with corresponding physical quantities. We shall see below that as far as a shift of the Fermi level is of interest the change of the apparent activation energy is a much less reliable measure than the change of the conductivity itself using a constant value for the prefactor, σ_0.

7.1.4 The Meyer–Neldel Rule of the Field Effect

We have shown in Sect. 3.5 that the temperature dependence of the source-drain current I_{SD} of a field effect transistor shows a Meyer–Neldel behaviour when the gate voltage is varied. We shall now demonstrate that a suitable modification of the statistical shift accounts for this fact. We shall not treat the field effect with all its complications (see Schumacher et al. [1988] for details) but consider the simplest case only: We have a MIS sample consisting of a semiconductor film of thickness l with a dielectric constant ε_s covered by an insulator of thickness d with a dielectric constant ε_d. The insulator and the interface to the semiconductor are assumed to be ideal in the sense that there are no states at the interface. For simplicity we shall also assume that there is no band bending for zero gate voltage, or in other terms that the flat-band voltage, V_{FB}, is zero. In this case the source-drain current, I_{SD}, equals the so-called flat-band current I_{FB}. We concentrate on nominally undoped films that are n-type. The flat-band current will be "activated" with an apparent activation energy E^*_σ and an intercept $I^*_{0,FB}$. A positive gate voltage, V_G, applied to the gate electrode gives rise to a potential $V(x)$ in the semiconductor ($x = 0$ is the interface plane) which induces a negative space charge $\varrho(x)$. The potential (which must be taken to be dependent on the space coordinate x) has the consequence that in thermal quasiequilibrium the carrier statistics is dominated by the electrochemical potential

$$E_F^{ec}(x) = E_F - q\,V(x) \tag{7.1.23}$$

which leads to a space charge density (with $q < 0$ for electrons)

$$\varrho(V(x)) = q \int_{-\infty}^{\infty} g(E)\,[f_F(E, E_F - qV(x), T) - f_F(E, E_F, T)]\,dE \tag{7.1.24}$$

which must obey Poisson's equation

$$V''(x) = \frac{-1}{\varepsilon_0\varepsilon_s}\,\varrho(V(x)) \tag{7.1.25}$$

and the corresponding boundary condition

$$V'(x = 0) = -\frac{\varepsilon_d}{\varepsilon_s\,d}\,[V_G - V(x = 0)] \quad . \tag{7.1.26}$$

The MIS structure can be considered as two capacitors in series (the insulator and the semiconductor). Since the capacitance of the semiconductor will depend somewhat on temperature the capacitance of the arrangement will not be temperature independent. Furthermore, as we are dealing with a spatially inhomogeneous system, there is no requirement of local charge conservation in the sense that $\varrho(V(x))$ should be independent of temperature. It is, therefore, the more intricate boundary condition (7.1.26) instead of the simpler requirement of charge conservation used so far which controls the temperature dependence of E_F^{ec}.

The potential $V(x)$ in the semiconductor, therefore, must depend on the temperature. This can be seen directly as follows: If we assume for the moment that $V(x)$ is temperature independent and raise the temperature we see from (7.1.24) that the negative charge density $\varrho(x)$ increases. From Poisson's law we see that V'' increases in contradiction to our current assumption that the potential $V(x)$ is independent of the temperature. This can be seen more clearly if we integrate Poisson's equation (neglecting $V'(x=l)$, for a justification of this nontrivial point see Grünewald et al. [1980]) and obtain

$$V'(x) = -\left[2\int_0^{V(x)} \frac{\varrho(V)}{\varepsilon_0\varepsilon_s}\,dV\right]^{1/2} \tag{7.1.27}$$

which shows directly that any increase of T that leads to an increase of the charge density must be accompanied by a decrease of V in order to fulfill the boundary condition at $x = 0$.

In order to illustrate the occurence of a MNR in the field effect current we shall expand the potential as a function of temperature as

$$V(x,T) = V(x,T=0) - \frac{1}{|q|}\,\gamma[V(x,T=0)]\cdot T \tag{7.1.28}$$

with

$$\gamma(V) = kC\,|q|\,V \quad . \tag{7.1.29}$$

For the current we can easily integrate

$$I_{\mathrm{SD}} = I_{\mathrm{FB}}\,\frac{1}{l}\int_0^l \exp[-q\beta V(x,T)]\,dx \tag{7.1.30}$$

which in the linear approximation gives

$$I_{\mathrm{SD}} = I_{\mathrm{FB}}\int_0^l \exp[-q(\beta - C)\,V(x,T=0)]\,dx \quad . \tag{7.1.31}$$

The Meyer–Neldel relation is more clearly seen if we introduce the effective depth, l_{eff}, of the highly conducting channel which for our purposes can be considered temperature independent (the "interface approximation"). We can then approximate the integral (7.1.31) by

$$I_{\mathrm{SD}} = I_{\mathrm{FB}}\,\frac{l_{\mathrm{eff}}}{l}\,\exp[-q(\beta - C)\,V(x=0\,,\,T=0)] \tag{7.1.32}$$

which shows that the source-drain current will appear to be activated with the "activation energy"

$$E_{\mathrm{a}}^{*} = E_{\sigma}^{*} + q\,V(x=0\,,\,T=0) \tag{7.1.33}$$

and the apparent prefactor

$$I_0^* = I_{0,\mathrm{FB}}^* \exp[qC\,V(x=0\,, T=0)] \tag{7.1.34}$$

which is exactly a Meyer–Neldel rule with $E_{\mathrm{MNR}}^* = 1/C$.

We see that the Meyer–Neldel rule for the field effect current is due to the temperature dependence of the electrochemical potential in a similar way as in the SCLC current. The fundamental difference between both situations is that in the SCLC case the charge density is uniform which is not true for the field effect. In the latter case the quasi-Fermi level must shift with rising temperature towards the equilibrium Fermi level in order to fulfill the boundary condition (7.1.26).

7.2 Temperature Dependence of the Electronic State Energies

As mentioned in Chap. 2 it is evident from optical absorption data that the states in the conduction bands move with respect to the valence band states as a function of temperature. This shift is described by

$$E_{\mathrm{g}}^{\mathrm{opt}}(T) = E_{\mathrm{g}}^{\mathrm{opt}}(0) - \gamma^{\mathrm{opt}}T \tag{7.2.1}$$

where the shift parameter depends slightly on the way how this shift is determined: using the shift of the photon energy at which a certain value of the absorption constant α is observed one finds

$$\gamma^{\mathrm{opt}} = (4.8 \pm 0.2)\cdot 10^{-4}\,\mathrm{eV/K} \tag{7.2.2a}$$

whereas the same evaluation for a constant value of $\alpha\hbar\omega$ gives

$$\gamma^{\mathrm{opt}} = (4.3 \pm 0.2)\cdot 10^{-4}\,\mathrm{eV/K} \tag{7.2.2b}$$

[Fritzsche, 1980]. We shall in the following use a mean of both values which in terms of Boltzmann's constant reads $\gamma^{\mathrm{opt}}/k = 5.4$.

From quantum mechanics we know that electronic states do not depend directly on the temperature. There may, however, be an indirect dependence on the temperature via the coupling to a quantum field because of temperature dependent occupation numbers of the states of the second field. For solids there are two mechanisms which contribute to a temperature dependence of electronic states:

i) With rising temperature one observes an expansion of the lattice which gives rise to a shift of the electronic energy levels.

This mechanism can be considered as rather independent from the actual occupancy of the electronic states. The thermal expansion is due to anharmonicity effects of the phonon system as a whole. The interaction with the electron system can be treated satisfactorily well by letting the lattice constant depend explicitely upon the temperature. Clearly, there must be an effect of the carrier occupancy in the different bands on the lattice expansion. It seems, however, to be a valid assumption to replace the actual carrier occupancy by the mean occupancy. This is in most cases equivalent to the assumption of explicitely temperature dependent electronic energies.

ii) The electron-phonon coupling contributes to the quasiparticle energy and gives rise to a temperature dependence due to the change of the mean phonon occupancy with temperature.

In this latter mechanism the mean energy of a given electronic state depends on the mean occupancy of the different phonon modes. If the electron-phonon interaction for states originating from the valence band is different from that for states from the conduction band one obtains a shift of the band gap.

This latter mechanism (which is considered to be the dominant mechanism for the red shift of the optical band gap in semiconductors) has been treated by Emin [1977] in the following model:
For the phonon system in the harmonic approximation the vibronic energy is given by

$$E_{\mathrm{vib}} = \sum_{\boldsymbol{q}} \hbar\omega_{\boldsymbol{q}}(n_{\boldsymbol{q}} + 1/2) \tag{7.2.3}$$

while the energy of some state in the conduction band is given by the sum of the bare electronic energy $E^0_{\boldsymbol{k}}$ and the interaction term with the phonons (N is the number of atoms in the system)

$$\tilde{E}_{\boldsymbol{k}} = E^0_{\boldsymbol{k}} - \frac{1}{kN} \sum_{\boldsymbol{q}} \hbar\omega_{\boldsymbol{q}}\, n_{\boldsymbol{q}}\, \gamma_{\boldsymbol{q}} \tag{7.2.4}$$

where the coupling constant $\gamma_{\boldsymbol{q}}$ may depend on the wavenumber $\boldsymbol{q}$ of the phonons. The system will for simplicity be described by the following classical Hamiltonian function

$$\begin{aligned} H(n_{\boldsymbol{k}}, n_{\boldsymbol{q}}) = &\sum_{\boldsymbol{k}} E^0_{\boldsymbol{k}} n_{\boldsymbol{k}} + \sum_{\boldsymbol{q}} \hbar\omega_{\boldsymbol{q}}(n_{\boldsymbol{q}} + 1/2) \\ &- \frac{1}{kN} \sum_{\boldsymbol{k}} \sum_{\boldsymbol{q}} \hbar\omega_{\boldsymbol{q}} \gamma_{\boldsymbol{q}} n_{\boldsymbol{k}} n_{\boldsymbol{q}} \quad . \end{aligned} \tag{7.2.5}$$

Since there is no electron-electron interaction we can evaluate the average occupation probability for an electron in state $\boldsymbol{k}$

$$n_{\boldsymbol{k}} = [\exp[\beta(\tilde{E}_{\boldsymbol{k}}(\langle n_{\boldsymbol{q}}\rangle) - E_{\mathrm{F}})] + 1]^{-1} \tag{7.2.6}$$

with

$$\langle n_{\boldsymbol{q}}\rangle = [\exp(\beta\hbar\omega_{\boldsymbol{q}}) - 1]^{-1} \quad . \tag{7.2.7}$$

We see that the occupation probability $\langle n_{\boldsymbol{k}}\rangle$ can be written as a Fermi–Dirac distribution function, however with some modified effective energy, $\tilde{E}_{\boldsymbol{k}}(\langle n_{\boldsymbol{q}}\rangle)$, and with the usual Bose–Einstein distribution function for the phonon occupation.

If we drop the zero point energy we can easily rearrange the terms in the Hamiltonian to obtain

$$H(n_{\boldsymbol{k}}, n_{\boldsymbol{q}}) = \sum_{\boldsymbol{k}} E^0_{\boldsymbol{k}} n_{\boldsymbol{k}} + \sum_{\boldsymbol{q}} \hbar\omega_{\boldsymbol{q}} \Big(1 - \frac{\gamma_{\boldsymbol{q}}}{kN} \sum_{\boldsymbol{k}} n_{\boldsymbol{k}}\Big) n_{\boldsymbol{q}} \tag{7.2.8}$$

which now suggests that the actual number of carriers in the conduction band leads to a softening of the phonon frequencies which in turn will lead to a modification of the mean phonon occupation numbers. We thus see that with interacting particles it is not easy to allocate the interaction energy to one kind of particles.

If the excitation of a carrier causes no immediate change of the phonon occupation numbers as is, e.g., assumed for optical excitations we obtain $\tilde{E}_{\boldsymbol{k}}(\langle n_{\boldsymbol{q}} \rangle)$ as the energy observed in absorption processes. This energy, however, is different from the energy

$$\langle E_{\boldsymbol{k}} \rangle = \langle H(\ldots, n_{\boldsymbol{k}} = 1, \ldots n_{\boldsymbol{q}}) \rangle - \langle H(\ldots, n_{\boldsymbol{k}} = 0, \ldots, n_{\boldsymbol{q}}) \rangle \tag{7.2.9}$$

which is the energy required to bring an extra electron into the conduction band in thermal equilibrium with the phonon system. A short calculation gives in the high temperature limit that $\langle E_{\boldsymbol{k}} \rangle = E^0_{\boldsymbol{k}}$ is the bare electronic energy (in the high temperature limit the mean energy in a given phonon mode is kT irrespective of the phonon energy).

For the dc conductivity we need the carrier density in equilibrium. This quantity which is given by (7.2.6) clearly has the activation energy $\tilde{E}_{\boldsymbol{k}}(\langle n_{\boldsymbol{q}} \rangle)$, i.e., the temperature dependent energy that also enters the optical gap. The situation is different, however, for the thermoelectric power. Since this quantity is connected to a temperature gradient we shall discuss the Peltier coefficient instead (note that Onsager's symmetry relation (2.5.2) which relates both coefficients holds irrespective of the electron-phonon coupling as long as the current densities are linear functions of the driving forces). In an idealized model we consider that carriers are excited from the Fermi level of one metal contact to some conducting states and drift through the sample until, at the other electrode, the carriers release the excitation energy. It is clear that in this idealized model it will be the energy $\langle E_{\boldsymbol{k}} \rangle$ which is transported by the carrier. As usual it is assumed that in thermal equilibrium effects like scattering, trapping and detrapping, recombination and reexcitation, etc., do not contribute to the heat current. We thus find that to a first approximation Emin's derivation is correct.

This treatment has been criticized by Butcher and Friedman [1977] and by Butcher [1984] (see also the reply by Emin [1985]) who concentrate on the electronic subsystem and, therefore, treat a system with explicitely temperature dependent electronic energies.

The controversy is not settled at the moment since the transport equations for the coupled electron-phonon system have not been treated properly for the case that there is scattering of the carriers. At least in the limits that there is no scattering of the carriers by phonons and in the case that the carriers are small polarons Emin's derivation seems to be correct.

If we assume that this result also holds in the case of amorphous semiconductors we expect the following effects (see [Overhof and Beyer, 1983] for details): we use an Einstein model for the phonons with $\hbar\omega_{\mathrm{E}}/k = 400\,\mathrm{K}$ and a constant value $\gamma_{\boldsymbol{q}} = \bar{\gamma}$. Prefactors and slopes are modified according to

$$\ln \sigma_0^*(\bar{\gamma}) = \ln \sigma_0^*(\bar{\gamma} = 0) + 0.9\,\bar{\gamma}/k \tag{7.2.10a}$$

$$E_\sigma^*(\bar{\gamma}) = E_\sigma^*(\bar{\gamma} = 0) + 0.3\,\bar{\gamma}\hbar\,\omega_0/k \tag{7.2.10b}$$

$$E_S^*(\bar{\gamma}) = E_S^*(\bar{\gamma} = 0) + 0.2\,\bar{\gamma}\hbar\,\omega_0/k \tag{7.2.10c}$$

$$A^*(\bar{\gamma}) = A^*(\bar{\gamma} = 0) + 0.2\,\bar{\gamma}/k \quad . \tag{7.2.10d}$$

As a result the intercept of the Q function will be modified according to

$$Q_0^*(\bar{\gamma}) = Q_0^*(\bar{\gamma} = 0) + 1.1\,\bar{\gamma}/k \tag{7.2.10e}$$

while

$$E_Q^*(\bar{\gamma}) = E_Q^*(\bar{\gamma} = 0) - 0.1\,\bar{\gamma}\hbar\,\omega_0/kT \quad . \tag{7.2.10f}$$

Thus far we have treated the case that the Fermi energy is in the valence band. For an amorphous semiconductor with the Fermi energy in about the middle of the pseudogap we must take into account that the lowering of the conduction band energy with respect to the states where the Fermi energy is situated enters in (7.2.6) only. If we, therefore, take an average value $\bar{\gamma} = \gamma^{\mathrm{opt}}/2$ we see that the most prominent effect will be an increase of the apparent prefactor of the conductivity, σ_0^*, by about a factor of 10 and an increase of Q_0^* by about 3.

In the following discussion the ambiguity of the energies for the interacting electron-phonon system could cause some confusion in particular if combined with the temperature dependence of, e.g., the Fermi energy. We shall, therefore, introduce temperature dependent electronic energies, $E_{\boldsymbol{k}}(T)$, which model the effect of the electron-phonon coupling. The effect of the electron-phonon coupling on the transport parameters can then be approximated by

$$\ln \sigma = \ln \sigma_0 - \beta[E_{\mathrm{C}}(T) - E_{\mathrm{F}}(T)] \tag{7.2.11a}$$

and

$$\frac{q}{k} S = \beta[E_{\mathrm{C}}(T) - E_{\mathrm{F}}(T)] + \bar{\gamma}/k + A \tag{7.2.11b}$$

which has the consequence that

$$Q_0^* = Q_0^*(\bar{\gamma} = 0) + \bar{\gamma}/k \tag{7.2.11c}$$

and

$$E_Q^* \cong E_Q(\gamma = 0) \quad . \tag{7.2.11d}$$

Note that in this notation the discrepancy between Emin on one side and Butcher and Friedman on the other side reduces to the additional term $\bar{\gamma}/k$ in (7.2.11b) and (7.2.11c) which according to the work of Butcher and Friedman is absent.

Strangely enough there is no experiment that could settle the controversy. In crystalline semiconductors, for which the above statements should be valid as well the thermoelectric power data are blurred by the phonon drag and, therefore, only the effect of the electron-phonon coupling on the conductivity can be measured which is not controversial. In liquid semiconductors the change of the band gap with temperature is a much more pronounced effect than in solids. Experimental data by Schmutzler and Hensel [1972] and by Freyland et al. [1974] on expanded fluid mercury well above the critical point do not indicate any effect of the electron-phonon coupling at all. But for liquid systems the band gap is probably primarily a function of the density and coordination of the atoms and, therefore, the electron-phonon coupling might be of minor importance for the temperature shift of the band gap in these systems.

8. The Transport Properties of Hydrogenated Amorphous Semiconductors

We are now in the position to collect the information presented in the previous chapters in order to compare our theoretical picture with the experimental results. To this aim we shall incorporate the transport properties of the homogeneous model as summarized in Sect. 4.8. We shall see that for doped semiconductors we cannot ignore the influence of the random potential discussed at some length in Chap. 6. Furthermore it is obvious from the previous chapter that the statistical shift of the Fermi energy gives rise to the different Meyer–Neldel rules while the red shift of the band gap has a less dramatic effect modifying the prefactors slightly.

We shall start with a discussion of hydrogenated amorphous silicon in which the scientific and also the technological interest is greatest. From the best estimate of the density of states distribution which varies with temperature, preparation, and doping level we derive the statistical shift which enables us to compare experimental and theoretical transport data.

We then proceed to study the amorphous germanium system for which many of the complications inherent in amorphous silicon are absent. It is, e.g., not unreasonable in the case of a-Ge : H to assume that the density of states distribution is essentially unaltered by temperature as well as by n-type and p-type doping. This simplifying assumption has the advantage that the predictions of the theory can be tested much more seriously against the experimental data.

8.1 Transport in Hydrogenated Amorphous Silicon

In this section we start with a detailed discussion of the density of states distribution in n-type a-Si : H and the corresponding Fermi level shift. We then proceed to discuss the dc conductivity, in particular the microscopic prefactor of the conductivity. This allows us to discuss the thermoelectric power and the Q data in a unified picture.

8.1.1 Density of States Distribution and the Statistical Shift of the Fermi Energy

The density of states distribution function of hydrogenated amorphous silicon has been the subject of much controversy in the last 15 years. The first field effect experiments [Spear and LeComber, 1972; Madan et al., 1976] resulted in a density of states function that for some time was considered valid by many groups. From deep level transient spectroscopy (DLTS) data by Cohen and coworkers [1980; see also Lang et al., 1982, and the review by Cohen, 1984, about the density of states distribution as derived from the various junction spectroscopy techniques] a density of states distribution emerged that in all essential features did not agree with the results of the field effect, while SCLC data resulted in still another density of states function. In the meantime a new density of states distribution function has been proposed by many groups [Street, 1985; Stutzmann and Stuke, 1983; Stuke, 1984; Jackson et al., 1985, Winer et al., 1988] that in many respects is similar to the results of the DLTS experiments and seems to describe most experiments.

This density of states distribution has the following features: There is a rather steep conduction band tail with a high density of states at the mobility edge. For undoped and for lightly doped material the density of deep midgap states is small consisting essentially of a rather sharp peak usually attributed to the dangling bonds. These states have a positive effective correlation energy, U_{eff}, which is of the order of 0.3 eV [Dersch et al., 1981; Stutzmann and Stuke, 1983]. The total density of these states depends on the doping level as is known from several independent experiments [Overhof and Beyer, 1979; Street and Biegelsen, 1980; Tanaka and Yamasaki, 1982; Jackson and Amer, 1982]. Street [1982] has proposed a model that can in principle account for the observed relation that the density of dangling bonds is proportional to the square root of the dopant concentration [see e.g., Stutzmann, 1986].

While the different experiments just mentioned seem to lead to a common density of states distribution there have appeared new experimental data that strongly advise a careful use of this distribution function: Pierz et al. [1987] find from their absorption data (obtained from photoconductivity experiments using the Constant Photocurrent Method) that there must be a deep minimum in the density of states distribution near the Fermi energy for p-doped samples. It follows that the energy of the dangling bond peak for p-type samples must be different from that for n-type samples. Kočka et al. [1987] obtain similar results which are also corroborated by their SCLC data. Winer et al. [1988] show from total yield photoemission data that for p-type samples the density of states near the Fermi energy is greatly reduced with respect to that of n-type samples. In the following we shall therefore restrict our discussion to n-type samples, where we may use the density of states distribution function proposed by Street [1985].

Figure 8.1 gives a quantitative picture of the density of states distribution for a-Si:H doped by phosphorus at different doping levels. For the valence and conduction band tail states we shall in the following assume that $U_{\text{eff}} \cong 0$ while for the dangling bonds we cannot ignore that the correlation energy is nonzero. The density of states for the doubly occupied D^- states in Fig. 8.1 must be regarded with some caution since these two-electron states do not fit into the one-electron picture. In particular these states can only be populated if the corresponding one-electron states D^0 are already occupied.

We shall in the following include the red shift of the optical band gap in the following way: we assume the states in the conduction band tail (and above) to

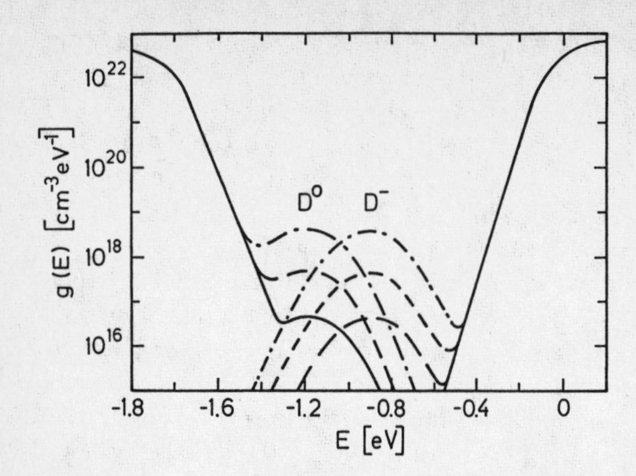

Fig. 8.1. Density of states distribution for a-Si:H at different doping levels. The density of dangling bonds varies from $10^{16}\,\mathrm{cm}^{-3}\,\mathrm{eV}^{-1}$ (*full line*) and 10^{17} (*dashed line*) to $10^{18}\,\mathrm{cm}^{-3}\,\mathrm{eV}^{-1}$ (*dashed-dotted line*)

shift rigidly with rising temperature towards the valence band tail states which we consider fixed. The shift will be considered linear with temperature at the rate $\gamma^{\mathrm{opt}} = 5.4\,k$ while the dangling bond states shift towards the conduction band states at half that rate.

In Fig. 8.2 we show a sequence of $E_{\mathrm{F}}(T)$ curves (with respect to the temperature dependent $E_{\mathrm{C}}(T)$) calculated for the density of states distributions of Fig. 8.1. At elevated temperatures the curves for a given dangling bond density but different position of $E_{\mathrm{F}}(T=0)$ tend to converge towards high temperature curves which strongly depend on the density of dangling bonds. Note that this high temperature part of the curves is rather meaningless because the density of states above some 410 K depends on the temperature itself [Kakalios and Street, 1986; Street et al., 1986; Street et al., 1987; Meaudre et al., 1988]. Due to the dynamics of the reaction between active and inactive donors with broken bonds the Fermi level becomes effectively pinned in the minimum between the conduction band and the D^- states.

This effective pinning of the Fermi level is only possible if the density of states changes with temperature. We see from Fig. 8.2 (where the density of states is temperature independent except for the temperature shift of the state energies described in Sect. 7.2) that the $E_{\mathrm{F}}(T)$ curves move through the minimum in the density of states without even a change of slope. It can be concluded quite generally that a deep minimum in the density of states distribution does *not* pin the Fermi energy. There is an exception, however: If the density of dangling bonds is very high and the material is undoped the Fermi level will be most effectively pinned in between the D^0 and the D^- states. This fact, however, is primarily not due to the minimum in the density of states but comes from the symmetry of the problem: there are exactly as many occupied states as unoccupied states and the density of occupied states is mirrored by the density of unoccupied states. For the highly asymmetric minimum of the density of states below the conduction band tail these arguments can of course not be used.

For the density of states distribution of Fig. 8.2 we plot the statistical shift parameter, γ_{F}^*, of the Fermi energy in Fig. 8.3 again taking the $T=0$ position of the Fermi energy as a free parameter. The shift parameter was obtained from the

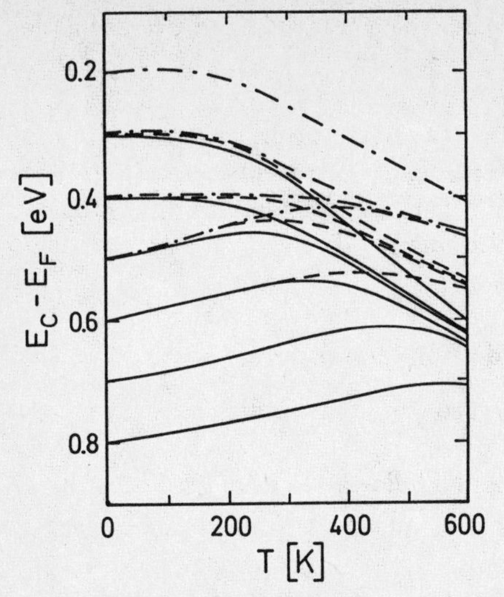

Fig. 8.2. Calculated Fermi level shift for the density of states models of Fig. 8.1

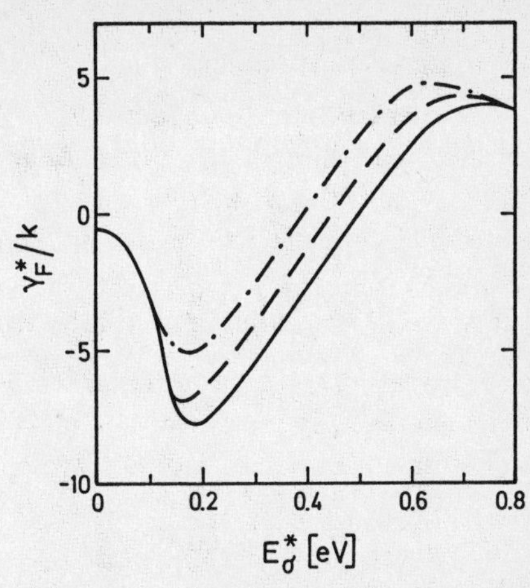

Fig. 8.3. Mean linear shift parameter γ_F^* calculated from the data of Fig. 8.2 in the 300 K < T < 400 K temperature range

statistical shift in the 300 K < T < 400 K temperature range. We mention that for the modelling of the Meyer–Neldel rule of the dc conductivity for a doping sequence we would have to alter the density of dangling bonds while changing the Fermi energy. The resulting curve would, therefore, interpolate between the dash-dotted line for the smaller values of E_σ^* and the full line at the higher values of E_σ^*, leading to $E_{\mathrm{MNR}} \cong 0.055$ eV.

One more technical point should be mentioned: Due to the finite effective correlation energy for the dangling bond states we have to use (7.1.13) for the evaluation of the neutrality condition. One might be tempted to replace this equation by the somewhat simpler expression (7.1.1) which seems to be a fairly good approximation to (7.1.13) if the density of states is replaced by some effective density of states. To be specific, we shall consider states that can be

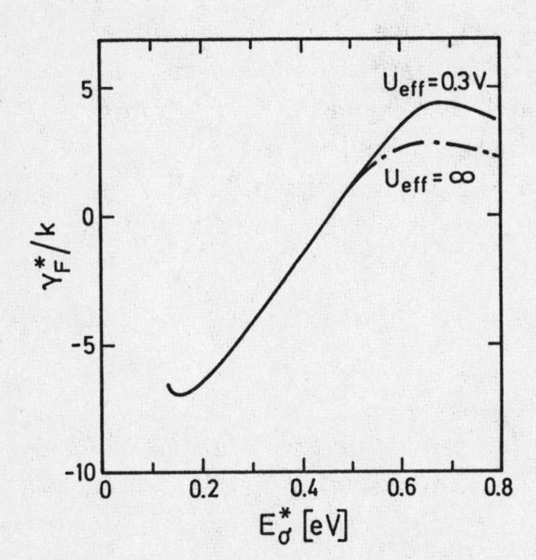

Fig. 8.4. Comparison of the mean linear shift parameters derived from the density of states model of Fig. 8.1 (*dashed line*) with a finite value for U_{eff} compared with data from an effective density of states with an infinite value of U_{eff}

singly occupied only (i.e., $U_{\text{eff}} = \infty$) but introduce two separate peaks in the density of states at the energies where we expect the D^0 and the D^- states, respectively. The resulting Fermi level shift is shown in Fig. 8.4 in comparison with the properly derived data. We see that except for the very large apparent activation energies there is virtually no difference between the two curves. There are significant differences, however, once the Fermi energy has moved into the dangling bond states: in this case the doubly occupied D^- states are absent if the corresponding D^0 states are not occupied.

8.1.2 The Microscopic Prefactor of the dc Conductivity

We have seen in Sect. 4.8 that for a homogeneous system rather independent of the particular shape of the band tail we can describe the electronic transport by our Standard Transport Model if we choose the following parameters:

$$\begin{aligned} \sigma_0 &= \sigma_M \cong 150\,\Omega^{-1}\,\text{cm}^{-1} \\ A &= 2.2 \\ Q_0 &= 7.2 \\ E_Q &= 0.06\,\text{eV} \quad . \end{aligned} \tag{8.1.1}$$

If we concentrate on the prefactor of the conductivity we see from (7.2.10a) that the effect of the shrinking band gap is properly described by a explicitely temperature-dependent band edge E_C in the activation term. We further find from (6.3.6) that the prefactor of the conductivity is practically not influenced by a long-ranged potential. We, therefore, conclude from the theory presented that the conductivity can be described by

$$\sigma(T) = \sigma_M \exp[-\beta(E_C(T) - E_F(T))] \tag{8.1.2}$$

and that this result should be valid for *all* disordered semiconductors.

In order to compare this value with experimental data we must find some examples where we know the temperature dependence of the activation factor. There are two experiments on quite differently prepared samples: Dersch [1983] has studied undoped samples that are irradiated by electrons in order to raise the spin density to $10^{18}\,\text{cm}^{-3}$. As mentioned above the Fermi level in this case will be fixed near midgap. The observed prefactor of the conductivity ($2300\,\Omega^{-1}\,\text{cm}^{-1}$) is, therefore, equal to the microscopic prefactor, σ_0, if we correct for the shrinking of the band gap. The result is

$$\sigma_0 = 150\,\Omega^{-1}\,\text{cm}^{-1} \tag{8.1.3}$$

which is just the prediction of the theory presented in Sect. 4.8.

The other case is the compensated sample already presented in Fig. 3.12 for which the thermoelectric power was exactly zero [Beyer et al., 1981]. This can happen only if

i) The contribution by electrons to the current is exactly equal to the contribution by holes in the temperature range studied.
ii) The prefactors of the conductivity for both carriers are exactly the same.

The observed prefactor of $4700\,\Omega^{-1}\,\mathrm{cm}^{-1}$ is within experimental error just twice the prefactor found by Dersch. We, therefore, find that there is excellent agreement between the theoretical value σ_{M} and the true experimental σ_0.

We should like to apply this result to a direct modeling of the Meyer–Neldel rule found in the Staebler–Wronski effect [Staebler and Wronski, 1977]: We simulate the SWE by the creation of additional dangling bonds caused by illumination [Dersch et al., 1981; Hauschildt et al., 1982; Weber et al., 1982; see also Jackson and Kakalios, 1988]. Upon formation these dangling bonds have one single electron only. In a n-type doped sample the dangling bond will capture a second electron and thereby move the Fermi energy towards midgap.

Our simulation of this effect is straightforward [Overhof, 1987]: In the state A the Fermi level is in the tail states and the total electron density, n_0, is calculated from the density of states distribution. The effect of illumination in our simulation consists in the creation of n^* additional dangling bonds. The resulting statistical shift of the Fermi energy $E_{\mathrm{F}}(T)$ for the system with $n_0 + n^*$ electrons combined with the value of σ_{M} gives the Meyer–Neldel rule shown as a full line in Fig. 8.5. This is compared with experimental data of Irsigler et al. [1983]. The agreement is better than could be expected from the simple model. Note that this density of states distribution readily accounts for σ_0^* values up to a few times $10^4\,\Omega^{-1}\,\mathrm{cm}^{-1}$. Larger values for σ_0^*, however, like those shown in Fig. 3.3 would require the use of a quite particular density of states distributions that would not be considered plausible at the moment.

In a recent paper [Overhof, 1987] an attempt has been made to explain the high temperature kink of the dc conductivity by the statistical shift of the Fermi energy alone using a stable density of states distribution. A detailed fit was possible only under the assumption of rather unrealistic σ_{M} values. The problem is satisfactorily solved by the observation of metastable densities of states.

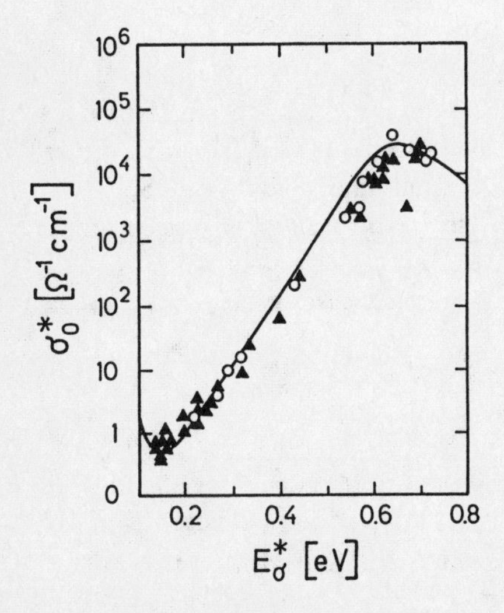

Fig. 8.5. Comparison of the experimental Meyer–Neldel rule data points obtained using the Staebler–Wronski effect data of Fig. 3.4 with the calculated curve

8.1.3 Determination of the Fermi Level Shift by Non-Transport Experiments

At this stage we should like to mention that there are two independent experimental methods which can be used to determine directly the Fermi level shift in a semiconductor. These are the Kelvin probe and the high resolution photoemission experiments.

In the Kelvin probe experiment [Kelvin, 1898; Johnson, 1958] a vibrating electrode is placed in front of the sample surface. The arrangement is equivalent to a capacitor with a time-dependent capacitance and will result in a time-dependent current if a voltage between the electrodes of the capacitor is applied. This ac current can be suppressed by means of a compensating dc voltage, V, which ensures that the Fermi energies of the sample surface and the vibrating electrode surface are equal ($\Delta E_{\mathrm{F}} = V$). In fact the Kelvin probe allows to measure the contact potential difference (CPD) between electrode and sample surface, in particular its temperature dependence.

With the same arrangement it is also possible to check if there is band bending at the sample surface by shining highly absorbed light onto the sample surface. This will give rise to a surface photovoltage which is a good measure of the band bending. The experiment by Foller et al. [1985] and by Siefert and de Rosny [1985; see also Foller et al., 1986] shows that for highly doped n-type samples the surface Fermi energy follows the bulk Fermi energy and agrees perfectly with the Fermi level derived from the conductivity data inverting (8.1.2). For undoped samples there is, however, a depletion layer at the sample surface which prohibits the study of the Fermi level shift.

Using the photoemission technique [Gruntz, 1981; see also Ley, 1984] the position of the Fermi energy at the sample surface can be determined directly with respect to the position of the valence states. The Fermi level shift derived by this experiment also equals the one derived from conductivity data on comparable samples inverting (8.1.2).

8.1.4 Thermoelectric Power and Q

We have shown that our theory describes the conductivity data perfectly well. A correct description of the thermoelectric power for the same samples is, therefore, equivalent to a correct description of the corresponding Q function. Since the latter does not contain the statistical shift of the Fermi energy it is much more convenient for our present discussion.

From Sect. 3.2 we remember that in practically all cases where data are reported the Q function can be represented by

$$Q(T) = Q_0^* - \beta E_Q^* \tag{8.1.4}$$

with a value for Q_0^* that is remarkably constant for all samples and close to 10 while E_Q^* is in the $0.05\,\mathrm{eV} < E_Q^* < 0.25\,\mathrm{eV}$ range. If we add up the contributions to Q_0^* from the homogeneous model, the long-ranged potential, and the shift of the band gap (assuming that Emin's treatment is correct) we obtain from (7.3.1) and from Sect. 4.8

$$Q^*_{0,\mathrm{theor}} = 7.2 + \Delta + \bar{\gamma}/k \tag{8.1.5}$$

with $\Delta = 1.8$ according to our resistor network calculation. In the random walk model the value for $\Delta = 3.5$ was somewhat higher. We are at present unable to decide which method is more accurate. In both cases the value for $Q^*_{0,\mathrm{theor}}$ is somewhat larger than the experimental value if we use $\bar{\gamma}/k = 2.7$ as appropriate for undoped samples. The situation is more complicated, however.

For undoped and for lightly n-doped samples one observes a value of $E^*_Q \cong 0.05\,\mathrm{eV}$ which is in the range covered by the homogeneous model alone. We, therefore, must assume that there is no significant effect of a long-ranged potential on the transport properties in these samples which is not too surprising as there will be few, if any, charged centers. As a consequence there will be no contribution of the Δ term in (8.1.5) while $\bar{\gamma} \cong \gamma^{\mathrm{opt}}/2$ will be appropriate. In this case the theoretical and the experimental values for Q^*_0 agree. For the moderately and highly doped samples the observed E^*_Q values suggest a strong influence of the long-ranged random potential (which must be present due to the charged dopants anyway) which results in $\Delta \cong 2-3$. The $\bar{\gamma}/k$ term in this case will be rather small because the Fermi level is already in the tail states.

While these considerations could satisfactorily explain the insensitivity to doping of the intercept Q^*_0 for n-type samples we run into difficulties for lightly p-doped samples and for heavily doped compensated samples. In these cases the Fermi level is within the dangling bonds and the large slope E^*_Q indicates a nonzero random potential. It is a puzzle that the observed values for Q^*_0 show such a little variation: while previously the observation of a universal intercept was taken as a welcome experimental proof that there is little change in the transport path upon doping etc., we now have difficulties to explain why there is so little variation.

Note that we only in part solve this problem if we abandon Emin's theory and follow Butcher and Friedman (see the discussion in Sect. 7.2). This brings us into the opposite difficulty that we do no longer obtain a large intercept Q^*_0 for the undoped samples. Clearly more experimental work would be highly welcome to clarify this point.

8.1.5 Drift Mobility Experiments for Doped and Compensated a-Si : H

The TOF experiment requires that the dielectric relaxation time is considerably larger than the transit time, a condition that cannot be fulfilled for the more conducting doped a-Si : H films. Here other experimental methods have been employed in order to obtain the drift mobility.

The Marburg group [Hoheisel and Fuhs, 1988] determine the drift mobility from a comparison of the steady state photoconductivity with the initial decay time of this photoconductivity if the light is switched off. The Chicago group [Chen and Fritzsche, 1983; Fritzsche and Chen, 1983; and Takada and Fritzsche,

1987] use the travelling wave technique of Adler et al. [1981] for highly doped samples. In this experiment a surface acoustic wave (SAW) is excited in a piezoelectric crystal. The normal component of the electric field of the SAW induces surface charge waves in the free surface and in the interface of the a-Si:H sample placed a few microns above the piezoelectric crystal. The longitudinal component of the same electric field caused by the SAW then leads to a surface current from which the mobility can be derived. Both experiments should lead to the same drift mobility as a (hypothetical) TOF experiment if nondegenerate statistics can be used (and if the sample properties are homogeneous). The experiments agree that the drift mobility does not change significantly upon doping.

In contrast, Street et al. [1988] determine a strong decrease of the drift mobility from their experiments which compare the dc conductivity with the density of electrons in the band tail states for bona fide identical samples. This latter quantity is obtained [Street and Zesch, 1984] by sweep-out experiments. Again the results should be identical to those of a TOF experiment but in fact seem to contradict the results mentioned above. The authors interpret their experiments assuming a shift of the mobility edge (by 80 meV for samples doped by $10^{-4} PH_3$) towards higher energies. They recognize that this shift is comparable to the width of the long-ranged potential for these samples as discussed in Chap 6.

From our discussion of the dc conductivity in the presence of a long-ranged potential it is clear that there will be no significant shift of the activation energy for the conductivity towards higher energies. In fact, we would predict a slight shift to lower energies instead. Furthermore such a shift would be readily observed in the experiments by Hoheisel and Fuhs and by Takada and Fritzsche.

We, therefore, shall interpret the experiment of Street et al. stressing the point that a long-ranged random potential must lead to an inhomogeneity on a mesoscopic scale. The density of occupied tail states n_{BT} in such a system is given by a convolution of the distribution function $p(V)$ of the random potential with the density of tail states $g_{\mathrm{T}}(E)$ (multiplied by the Fermi distribution function)

$$n_{\mathrm{BT}} = \int_{-\infty}^{\infty} dE \int_{-\infty}^{\infty} dV\, p(V)\, g_{\mathrm{T}}(E+V)\, f_{\mathrm{F}}(E, E_{\mathrm{F}}, T) \quad . \tag{8.1.6}$$

For a rapidly rising tail density of states this convolution leads to a drastic increase of n_{BT} if the width, δ, of the random potential is increased. Detailed calculations [Overhof and Silver, 1989] show that this effect quantitatively explains the decrease of the drift mobility derived from the ratio of the dc conductivity to the density of band tail states.

For the travelling wave experiment of the Chicago group there is no theory that can take into account the threedimensional inhomogeneity due to the random potential. In this experiment, however, essentially a dc current is monitored, since the wavelength and the penetration depth of the SAW field are large compared to both $l_{\max}$, the length scale of the random potential, and to the sample thickness. We would, therefore not expect any strong dependence of the results on δ.

It is doubtful that in the photocurrent decay experiment of the Marburg group [Hoheisel and Fuhs, 1988] the long-ranged potential is of any importance at all. In particular for the moderately doped samples the high density (10^{16} cm^{-3}) of photoexcited carriers will screen any random potential. If for the highly doped samples there is a residual long-ranged random potential its influence on the resulting drift mobility will be small since both steady state photoconductivity and initial decay time constant are determined by the kinetics of the carriers at energies near the centroid of the transport path at E_t.

For compensated samples the drift mobility can be determined by the TOF experiment [Marshall et al., 1985; Spear et al., 1987]. It is found that the drift mobility decreases with increasing doping level and that the apparent activation energy of the drift mobility increases. While for the undoped sample the drift mobility derived from the transit time is strongly field dependent, this "dispersive" behaviour becomes less prominent with increasing doping level although the shape of the transients become more dispersive. We have not treated the case that both the long-ranged potential (which will be large for compensated samples for which the density of midgap states is small [Marshall et al., 1984]) and multiple trapping are combined to cause dispersive current transients. But we would like to speculate that the increase of the long-ranged potential upon doping for compensated samples leads to an increased apparent activation energy, to increased dispersive current transients and at the same time to a diminished dispersion as derived from the field dependence of the drift mobility.

8.1.6 Interpretation of Transport Data in Terms of the Model of Two Separate Conduction Channels

From the beginning of the study of amorphous silicon the difference between the slopes of the Arrhenius plot for the thermoelectric power and that for the dc conductivity has intrigued many researchers. The most popular solution to the problem has already been mentioned in Sect. 2.8.1, the model of two conduction channels. This model has been used extensively by many groups in order to parametrize the transport properties and the earlier data for a-Ge:H [Jones et al., 1977] and for a-As [Mytilineou and Davis, 1977] seemed to confirm the model. In most cases, however, the parameters of this model lead to severe inconsistencies. Ghiassy et al., [1985], e.g., are forced to postulate extremely high values for the heat of transport term, A, of the thermoelectric power due to hopping. If one tries to avoid these unrealistic values one has to assume that the microscopic prefactor of the conductivity, σ_0, for transport in the extended states varies by several orders of magnitude. We, therefore, would like to discard this simple model and prefer the theory of the previous chapters which also treats transport in extended states and hopping in tails, however, as two modes of transport that are highly interrelated.

8.2 Transport in Hydrogenated Amorphous Germanium

While technological interest has stimulated a vast amount of research in the a-Si:H field a material that in many respects is similar, a-Ge:H, has received comparably little attention, although we shall demonstrate that the transport properties of this material can be more easily understood than those of silicon. As can be seen from Fig. 3.13 the transport properties of a-Ge:H are quite unusual: for highly n-type and p-type doped material the thermoelectric power is quite large but virtually constant for a temperature range of some 200 K. Furthermore, there is a systematic difference between the Q values for p-type and those for n-type a-Ge:H which was not observed for a-Si:H. We shall see that these unusual features can be easily explained using the theoretical tools derived above.

8.2.1 Density of States and Spin Density

From the extensive work of the Marburg group [Stutzmann et al., 1983; Stutzmann and Stuke, 1983] the spin densities for three different groups of states in a-Ge:H as a function of the doping level are known. The appropriate Fermi level positions can be derived from transport studies on bona fide similar samples [Hauschildt, 1982]. Three different paramagnetic centers are detected which differ by their respective g values, linewidth, and saturation characteristics. Figure 8.6 shows the spin densities which are readily interpreted in terms of a valence band tail, a conduction band tail, and the dangling bond peak in the center of the pseudogap. We observe that the spin density for the states at midgap is remarkably higher than in the case of a-Si:H and that these spin density levels can be interpreted in terms of a rather broad density of states peak that does not change with doping.

In sharp contrast to the case of a-Si:H we shall, therefore, assume for a-Ge:H that the density of states distribution is always the same irrespective of the doping level (this may not be strictly true, however, but for the calculation of the statistical shift of the Fermi energy this assumption is sufficiently accurate).

The effective correlation energy, U_{eff}, for the different paramagnetic centers can be estimated from the temperature dependence of the observed spin densities. We follow here the values given by Stutzmann and Stuke [1983] in order to convert

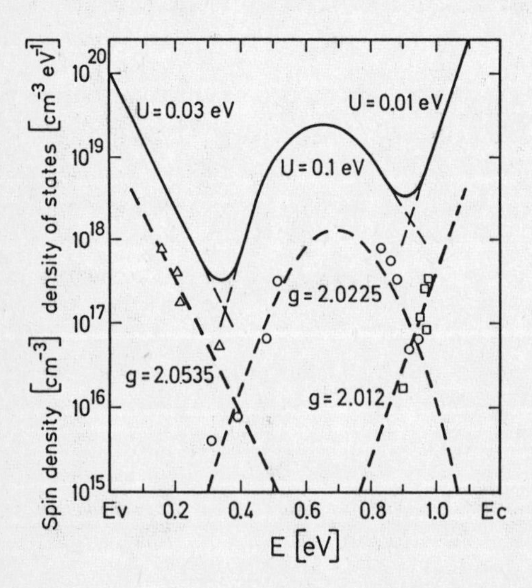

Fig. 8.6. Spin density for undoped and doped a-Ge:H films as a function of the Fermi energy (after [Stutzmann et al., 1982]). The density of states distribution function is calculated using the U_{eff} data indicated in the figure

the spin densities into a density of states distribution. In principle that requires a deconvolution which for the present case is quite simple and can be replaced by the inverse convolution using a trial density of states. The density of states distribution function thus obtained is also shown in Fig. 8.6. The most prominent feature is a broad maximum of the density of states (the dangling bond peak) in the center of the pseudogap with a width that considerably exceeds the effective correlation energy, U_{eff}.

This has the important consequence that for any position of the Fermi energy there is always a sizable fraction of the dangling bonds in a charged state: Heavy p-type doping leaves the dangling bond states as D^+, heavy n-type doping leads to a majority of D^-, and for undoped samples we have D^0, D^-, and D^+ states of comparable density. According to (6.2.4) we must expect a long-ranged random potential for which in all cases the rms amplitude is rather independent of the actual position of the Fermi energy. This result is exactly what we anticipate from the slope of the Q values of Fig. 3.14.

8.2.2 The Prefactor of the Conductivity and Q_0^*

We have no reason to assume that the prefactor of the conductivity, σ_0, should be different from σ_M (8.1.1) which is $150\,\Omega^{-1}\,\text{cm}^{-1}$ or thereabouts. The significant difference between the Q_0^* values for electrons (~ 10) and holes (~ 12) must then have a different origin. We suggest instead that the temperature dependent shift of the dangling bond peak is asymmetric with respect to the band edges: In a-Ge:H we have $\gamma^{\text{opt}} \simeq 4k$. If we assume that for the dangling bond states $\gamma^{\text{db}} \simeq 3k$ with respect to the valence band we find from (8.1.5) that for electrons ($\bar{\gamma} \leq \gamma^{\text{opt}} - \gamma^{\text{db}}$) $Q_0^* \simeq 10$ while for holes ($\bar{\gamma} \leq \gamma^{\text{db}}$) we obtain $Q_0^* \simeq 12$. We also expect that the intercept Q_0^* should be the largest for the undoped and lightly doped samples and smaller for the more heavily doped samples, in agreement with the experimental observation.

8.2.3 The Statistical Shift of the Fermi Energy and the Temperature Dependence of the Transport Properties

From the density of states distribution and the knowledge of the long-ranged potential it is now straightforward to calculate the transport properties of doped a-Ge:H explicitely with the $T = 0$ position of the Fermi energy as a free parameter. Figure 8.7 shows a plot of the Fermi level shift as a function of temperature. The energies are plotted with respect to a fixed valence band tail, the position of the conduction band mobility edge is indicated by the heavy double-line. Since we have assumed a severe asymmetry of the shift for the dangling bond peak we find the same asymmetry reflected in the $E_F(T)$-curves. This causes a rather rapid shift of the Fermi energy for undoped samples towards the valence band that must ultimately lead to ambipolar conduction. This is seen in the experimental data for the thermoelectric power (Fig. 3.13), where the rapid decrease for the undoped film above 400 K must be due to a contribution from holes (which has already been observed by Beyer and Stuke [1974] for evaporated germanium films). The same effect is seen even more pronounced in the Q data of Fig. 3.14.

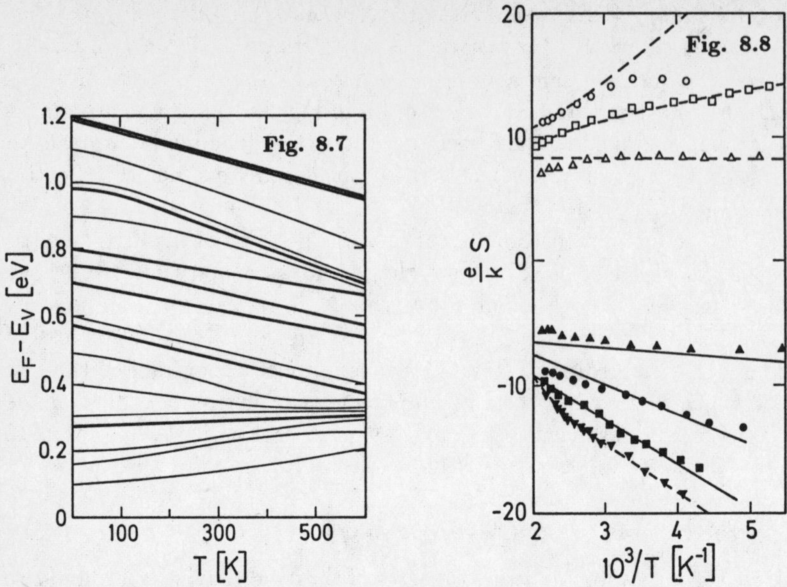

Fig. 8.7. Fermi level as a function of temperature for the density of states distribution of a-Ge : H given in Fig. 8.6 [Overhof, 1985]

Fig. 8.8. Thermoelectric power data for n-type (*full symbols*) and p-type (*open symbols*) a-Ge : H (after [Hauschildt, 1982]) compared with curves calculated from the theory presented [Overhof, 1985]

A corresponding transition to ambipolar conduction is not expected for p-type material (in the temperature range under consideration). Here more detailed experiments for undoped and lightly doped samples would be most welcome to clarify the asymmetry of the shift postulated for the dangling bond peak.

We present in Fig. 8.8 a comparison between the experimental [Hauschildt et al., 1982] and the calculated [Overhof, 1985] data for the thermoelectric power. Note that the calculated shift of the Fermi level quite naturally reproduces (in fact exaggerates) the strange temperature independence of the data for highly doped samples of both polarities.

9. Concluding Remarks

In this book we have presented a theory for the dc electronic transport in amorphous semiconductors that allows for a detailed and consistent interpretation of the various equilibrium and quasiequilibrium transport experiments reported for hydrogenated silicon and germanium.

We have shown that even for the simplest case of a disordered semiconductor which is homogeneous on a mesoscopic scale the transport mechanism is quite complex: The more intuitive picture of a distinct mobility edge separating extended (conducting) states from localized (essentially nonconducting) states cannot hold for such a semiconductor, in particular because at nonzero temperatures the interaction with phonons must be taken into consideration. We find that there is both a tunneling-type transport contribution, characteristic for conduction in extended states and also a hopping-type transport contribution, and that both contributions are highly inter-related, the underlying mechanism being phonon-induced delocalization. The ratio of the respective contributions to the total differential conductivity varies as a function of energy and depends somewhat on the particular model assumptions. The quantities that can be measured, however, are always integrals over the differential conductivities multiplied by the appropriate Boltzmann factor. We are primarily interested in the microscopic prefactor σ_0 which turns out to be surprisingly insensitive to the particular values of the model parameters with the single exception of one microscopic length, the average distance of sites in a tight binding model for the band tail. This length can be assumed to be quite independent from the chemistry, preparation conditions, and thermal (or illumination, etc.) history of the sample. Theoretical estimates of it as well as information from particular transport experiments give $\sigma_0 = 150\,\Omega^{-1}\,\mathrm{cm}^{-1}$. The most important result of our theory, however, is the uniqueness of the microscopic prefactor: It is difficult to find reasonable model parameters by which this value can be altered by more than a factor of 2. The other quantity that can be determined directly from the microscopic theory is E_Q^*, which turns out to be nonzero as a result of the rather wide energy range that contributes significantly to the conductivity. The actual value of E_Q^* is of course somewhat more model dependent. It turns out, however, that under reasonable assumptions we always have $E_Q^* < 0.1\,\mathrm{eV}$.

A homogeneous model on a mesoscopic scale cannot hold for moderately or highly doped amorphous semiconductors if the density of midgap states is considered to be small. In this case we must have a long-ranged random potential of electrostatic nature. A similar potential is expected for light-soaked samples and samples for which the dangling bond peak is very large, as will be the case for a-Si : H samples annealed at extreme temperatures and also for a-Ge : H. We have shown by extensive resistor network calculations and also by Monte Carlo random walk studies that to first order the random potential leads to an increase of E_Q^*. The random potential reduces the averaged Peltier heat, leaving all other transport parameters essentially unchanged.

These latter calculations also do not indicate that there might be a larger spread of the observed prefactors for the conductivity, σ_0^*, for which the experiments always give a tremendous variation. The apparent prefactor depends, in the form of the Meyer–Neldel rule (3.1.1)

$$\ln(\sigma_0^* \, \Omega \, \mathrm{cm}) = B_{\mathrm{MNR}} + E_\sigma^* / E_{\mathrm{MNR}} \quad ,$$

on the observed activation energies E_σ^*. We conclude that the various Meyer–Neldel rules derived for the different experiments are primarily due to the statistical shift of the Fermi energy. For this statement we have presented the following evidence:

i) From any reasonable model for the density of states distribution one can easily calculate the statistical shift, which in all cases results in a Meyer–Neldel rule. There is no need to rely on a particular model for the density of states. One can base the calculation on the more realistic models proposed recently and thereby obtain the parameters observed experimentally.

ii) The intercept of the experimentally observed Q function Q_0^* is virtually independent of any sample parameter and amounts to 10 ± 1 in all cases. It is difficult to imagine that the observed heat of transport term A^* and the apparent prefactor of the conductivity $\ln(\sigma_0^*)$ should always add up to 10 if the variation of the individual quantities is caused by something other than the statistical shift.

iii) The various experiments that show a Meyer–Neldel rule for a single sample for which the transport properties can be altered in a reversible manner (field effect, space-charge limited current, Staebler–Wronsky effect) can most naturally be interpreted in terms of the statistical shift of the Fermi level (or quasi-Fermi energy or electrochemical potential, respectively). We do not know of any other consistent interpretation of the different Meyer–Neldel rules.

We therefore see that the most accurate way to determine the position of the Fermi energy from transport data is the inversion of (8.1.2)

$$E_{\mathrm{C}}(T) - E_{\mathrm{F}}(T) = kT \, \ln(\sigma_0 / \sigma(T)) \quad ,$$

which can be evaluated without any extrapolation for the temperature of interest. We warn against the use of E_σ^*, which seems to enjoy an enormous popularity among experimentalists in spite of the ubiquitous Meyer–Neldel rule. The error introduced in the estimate of E_F due to this unjustifiable procedure can well exceed 0.2 eV. It is sometimes claimed that the apparent activation energy should be used to determine the position of the Fermi energy because activation energies can be "directly determined from the experiment". It is clear from the above that this argument is invalid unless there is a theory that can explain the spread of the apparent prefactors σ_0^* by a corresponding spread of the microscopic prefactors.

There is no reason why our present theory should not be applicable to other disordered semiconductors. We have mentioned in Sect. 3.2 that for the chalcogenide glasses the Q function is very similar to that observed for the tetrahedrally bonded hydrogenated semiconductors. Unfortunately there are very few experiments where the thermoelectric power has been measured over a significant temperature range.

Our theory describes the dc transport experiments reported for a-Si:H and a-Ge:H, making several predictions that can be checked experimentally. It would, e.g., be of greatest interest if one could find disordered systems for which the slope E_Q^* is zero or for which the values of the intercepts Q_0^* significantly deviate from our value of 10 ± 1.

As an example of non-equilibrium transport we have also considered the time-of-flight experiments. It was shown that the long-ranged potential can provide a natural explanation for the anomalous transients which are usually interpreted in terms of hopping in tails or multiple trapping, or both. A more complete theory of the time-of-flight experiment in amorphous semiconductors, treating hopping, transport in extended states with trapping, and the semiclassical long-ranged potential on an equal footing does not yet exist, but could be worked out along the lines presented here for equilibrium transport.

A cumbersome problem in the amorphous semiconductor field is the correlation of data obtained on bona fide comparable samples in different experiments. We believe that it could be worthwhile performing the different experiments on a single set of samples. At present most experimentalists optimize the sample preparation in such a way that the experimental technique used gives the best results, i.e. best resolution, etc. For the determination of the electron spin resonance, e.g., the sample is treated in a way that ensures a high spin density, while for most other experiments one tries to prepare samples with a minimum defect density.

For a comparison of experimental data with our present theory it would be of utmost interest if one could verify (or falsify) the correlation predicted in Chap. 6 between the value of E_Q^* and the characteristic energies observed in the ac conductivity, the time-of-flight experiment, and the superlinear current experiment, which is commonly interpreted as space-charge limited current and for which we propose an alternate interpretation.

From the experimental point of view several questions are open and require more work to be done. The interaction between dopant atoms, the hydrogen atoms, and defects like dangling bonds as a function of the Fermi energy seems to be a problem that should find a solution in the near future. This hopefully could also solve the question how in detail the density of states distribution is altered upon doping. Another problem is connected with the time-of-flight experiments, where there are several discrepancies between the experimental results of different groups. In particular the very low temperature results reported recently by the Dundee group do not seem to be understandable.

From the theoretical side much work is needed in order to come to a satisfactory solution of the open problems. First of all one would like to have a more detailed picture of the electronic states in the tails. If it were possible to include electron-phonon scattering into a transport theory, one could also investigate the interesting problem of a system in the limit of small disorder. A major task would be to incorporate the static aspects of the electron-phonon interaction in order to solve the problems connected with small polaron formation, for which we have provided qualitative arguments only.

At present the most unsatisfactory problem in the physics of electronic transport phenomena observed in amorphous semiconductors is the sign of the Hall effect. It seems that a sign anomaly (with respect to all other experiments that provide information about the sign of the predominant carriers) is always present, but again the experimental data are relatively scarce. No theory except for the small polaron model can account for this sign anomaly; therefore does this mean that we would be better off dealing with small polarons to begin with? The abundance of experiments presented in this book can evidently be interpreted satisfactorily by a theory that does not rely on small polarons, but we must admit that we have no solution for the Hall effect at hand.

References

Chapter 1

Ambegaokar, V., B.I. Halperin, J.S. Langer, 1971: Phys. Rev. B4, 2612
Anderson, P.W., 1958: Phys. Rev. **109**, 1492
Anderson, P.W., 1972: Nature, Phys. Sci. (London) **235**, 163
Ando, T., 1982: in *Anderson Localization*, Springer Ser., Solid-State Sci., Vol. **39**, ed. by Y. Nagaoka, H. Fukuyama (Springer, Heidelberg, New York, Berlin) p. 176
Aoki, H., 1982: Physica **114A**, 538
Aoki, H., 1983: J. Phys. C**16**, L205
Bányai, L., 1964: in *Physique de Semiconducteurs*, ed. by M. Hulin (Dunod, Paris) p. 417
Barna, Á., P.B. Barna, Z. Bodó, J.F. Pócza, I. Pozsgai, G. Radnóczi, 1974: in *Proc. 5th Int. Conf. on Physics of Amorphous and Liquid Semiconductors*, ed. by J. Stuke, W. Brenig (Taylor and Francis, London) p. 109
Beaglehole, D., M. Zavetova, 1970: J. Non-Cryst. Sol. **4**, 272
Böttger, H., V.V. Bryksin, 1985: *Hopping Conduction in Solids* (Akademie Verlag, Berlin)
Butcher, P.N., J.A. McInnes, 1978: Phil. Mag. **B37**, 249
Carius, R., W. Fuhs, 1985: J. Non-Cryst. Sol. **77/78**, 659
Chittick, R.C., J.H. Alexander, H.F. Sterling, 1969: J. Electrochem. Soc. **116**, 77
Chomette, A., B. Deveaud, A. Regreny, G. Bastard, 1986: Phys. Rev. Lett. **57**, 1464
Cohen, M.H., E.N. Economou, C.M. Soukoulis, 1983: J. Non-Cryst. Sol. **59/60**, 15; Phys. Rev. Lett. **51**, 1202
Cohen, M.H., E.N. Economou, C.M. Soukoulis, 1984: Phys. Rev. B**29**, 4496 and 4500
Cohen, M.H., H. Fritzsche, S.R. Ovshinsky, 1969: Phys. Rev. Lett. **22**, 1065
Dersch, H., L. Schweitzer, J. Stuke, 1983: Phys. Rev. B**28**, 4678
Donovan, T.M., W.E. Spicer, 1968: Phys. Rev. Lett. **21**, 1572
Goryunova, N.A., B.T. Kolomiets, 1955: Zh. tekh. Fiz. **25**, 984 and 2069
Hall, G.G., 1952: Phil. Mag. **43**, 338
Hall, G.G., 1958: Phil. Mag. **3**, 429
Hass, G., 1947: Phys. Rev. **72**, 174
Imry, Y., 1980: Phys. Rev. Lett. **44**, 469
Ioffe, A.F., A.R. Regel, 1960: Progr. Semicond. **4**, 237
Kenkre, V.M., P. Reineker, 1982: *Exciton Dynamics in Molecular Crystals and Aggregates*, Springer Tracts in Mod. Phys., Vol. **94** (Springer, Heidelberg, New York, Berlin)
Klitzing, K.v., G. Dorda, M. Pepper, 1980: Phys. Rev. Lett. **45**, 494
Ley, L., 1984: in *Semiconductors and Semimetals* **21c**, 385, ed. by J. Pankove (Acad. Press, New York)
Löhneisen, H.v., 1983: J. Non-Cryst. Sol. **59/60**, 1087
Maschke, K., H. Overhof, P. Thomas, 1974: phys. stat. sol. (b) **62**, 113
Matyàš, M., J. Kočka, B. Velicky (eds.), 1987: *Amorphous and Liquid Semiconductors*, J. Non-Cryst. Sol. **97/98**
McInnes, J.A., P.N. Butcher, J.D. Clark, 1980: Phil. Mag. B**41**, 1
Meyer, W., H. Neldel, 1937: Z. Techn. Phys. **12**, 588
Mott, N.F., 1967: Adv. Phys. **16**, 49
Mott, N.F., 1969: Phil. Mag. **19**, 835

Mott, N.F., E.A. Davis, 1971 and 1979: *Electronic Processes in Non-Crystalline Materials* (Clarendon Press, Oxford)
Müller, H., P. Thomas, 1983: Phys. Rev. Lett. **51**, 702
Müller, H., P. Thomas, 1984: J. Phys. C**17**, 5337
Noll, G., E.O. Göbel, 1987: J. Non-Cryst. Sol. **97/98**, 141
Overhof, H., W. Beyer, 1981: Phil. Mag. **B43**, 433
Ovshinsky, S.R., 1968: Phys. Rev. Lett. **21**, 1450
Phillips, J.C., 1979: Phys. Rev. Lett. **42**, 1151
Polk, D.E., 1971: J. Non-Cryst. Sol. **5**, 365
Pollak, M., 1972: J. Non-Cryst. Sol. **11**, 1
Pollak, M., A.L. Efros (eds.), 1985: *Electron-Electron Interactions in Disordered Systems* (North Holland, Amsterdam)
Sabine, R., 1878: Phil. Mag. S.5, Vol. **5**, 401
Schweitzer, L., B. Kramer, A. McKinnon, 1985: Z. Phys. **B59**, 379
Soukoulis, C.M., E.N. Economou, 1984: Phys. Rev. Lett. **52**, 565
Spear, W.E., P.G. LeComber, 1975: Solid State Commun. **17**, 1193
Sterling, H.F., R.C.G. Swann, 1965: Solid State Electr. **8**, 653
Stuke, J., 1953: Z. Phys. **134**, 194
Tauc, J., R. Grigorovici, A. Vancu, 1966: phys. stat. sol. **15**, 627
Thouless, D.J., 1977: Phys. Rev. Lett. **39**, 1167
Triska, A., D. Dennison, H. Fritzsche, 1975: Bull. Am. Phys. Soc. **20**, 392
Weaire, D., M.F. Thorpe, 1971: Phys. Rev. B**4**, 2508 and 3518
Weiser, G., U. Dersch, P. Thomas, 1988: Phil. Mag. **B57**, 721
Zeller, R.C., R.O. Pohl, 1971: Phys. Rev. B**4**, 2029

Chapter 2

Abrahams, E., P.W. Anderson, D.C. Licciardello, T.V. Ramakrishnan, 1979: Phys. Rev. Lett. **42**, 673
Abou-Chacra, R., D.J. Thouless, 1974: J. Phys. **C7**, 65
Alben, R., D. Weaire, J.E. Smith Jr., M.H. Brodsky, 1975: Phys. Rev. **B11**, 2271
Allen, P.B., M. Cardona, 1981: Phys. Rev. B**23**, 1495; Phys. Rev. B**24**, 7479
Ambegaokar, V., B.I. Halperin, J.S. Langer, 1971: Phys. Rev. B**4**, 2612
Anderson, P.W., 1958: Phys. Rev. **109**, 1492
Anderson, P.W., 1972: Nature Phys. Sci. (London) **235**, 163
Bányai, L., 1964: in *Physique de Semiconducteurs*, ed. by M. Hulin (Dunod, Paris) p. 417
Barna, Á., P.B. Barna, Z. Bodó, J.F. Pócza, I. Poszgai, R. Radnóczi, 1974: in *Proc. 5th Int. Conf. on Physics of Amorphous and Liquid Semic.*, ed. by J. Stuke, W. Brenig (Taylor and Francis, London) p. 109
Barna, Á., P.B. Barna, G. Radnóczi, H. Sugawara, P. Thomas, 1976: in *Structure and Excitations in Amorphous Solids*, ed. by G. Lucovsky, F.G. Galeener (American Institute of Physics, New York) p. 199
Barna, Á., P.B. Barna, G. Radnóczi, L. Tòth, P. Thomas, 1977: phys. stat. sol. (a) **41**, 81
Bar-Yam, Y., D. Adler, J.D. Joannopoulos, 1986: Phys. Rev. Lett. **57**, 467
Bar-Yam, Y., J.D. Joannopoulos, 1986: Phys. Rev. Lett. **56**, 1213; Phys. Rev. Lett. **56**, 2203
Baum, J., K.K. Gleason, A. Pines, A.N. Garroway, J.A. Reimer, 1986: Phys. Rev. Lett. **56**, 1377
Beaglehole, D., M. Zavetova, 1970: J. Non-Cryst. Sol. **4**, 272
Belitz, D., W. Götze, 1983: Phys. Rev. B**28**, 5445
Beyer, W., R. Fischer, 1977: Appl. Phys. Lett. **31**, 850
Beyer, W., H. Overhof, 1984: in *Semiconductors and Semimetals*, ed. by R.K. Willardson, A.C. Beer, Vol. **21c** (Academic Press, New York) p. 257
Beyer, W., H. Wagner, H. Mell, 1981: Solid State Commun. **39**, 375
Böttger, H., V.V. Bryksin, 1976: phys. stat. sol. (b) **78**, 9 and 415
Böttger, H., V.V. Bryksin, 1985: *Hopping Conduction in Solids* (Akademie Verlag, Berlin)

Brenig, W., G.H. Döhler, P. Wölfle, 1971: Z. Phys. **246**, 1
Brenig, W., G.H. Döhler, P. Wölfle, 1973: Z. Phys. **258**, 381
Brodsky, M.H., R.S. Title, 1969: Phys. Rev. Lett. **23**, 581
Brodsky, M.H., R.S. Title, K. Weiser, G.D. Pettit, 1970: Phys. Rev. **B1**, 2632
Brodsky, M.H., M. Cardona, J.J. Cuomo, 1977: Phys. Rev. **B16**, 3556
Brust, D., 1969: Phys. Rev. Lett. **23**, 1232; Phys. Rev. **186**, 768
Busch, G., H. Labhardt, 1946: Helv. Phys. Acta **19**, 463
Butcher, P.N., 1974: J. Phys. **C7**, 2645
Butcher, P.N., 1985: in *Amorphous Solids and the Liquid State*, ed. by N. Marsh, R.A. Street, M. Tosi (Plenum Press, New York) p. 311
Cardona, M., 1983: phys. stat. sol. (b) **118**, 463
Chaikin, P., G. Beni, 1976: Phys. Rev. **B13**, 647
Cohen, M.H., H. Fritzsche, S.R. Ovshinski, 1969: Phys. Rev. Lett. **22**, 1065
Connell, G.A.N., R.J. Temkin, 1974: Phys. Rev. **B9**, 5323
Conwell, E.M., 1956: Phys. Rev. **103**, 51
Cutler, M., N.F. Mott, 1969: Phys. Rev. **181**, 1336
Dersch, U., M. Grünewald, H. Overhof, P. Thomas, 1987: J. Phys. **C20**, 121
Döhler, G.H., 1979: Phys. Rev. **B19**, 2083
Donovan, T.M., 1970: Ph.D. Thesis, Stanford (unpublished)
Donovan, T.M., E.J. Ashley, W.E. Spicer, 1970: Phys. Lett. **A 32**, 85
Donovan, T.M., K. Heinemann, 1971: Phys. Rev. Lett. **27**, 1794
Donovan, T.M., W.E. Spicer, 1968: Phys. Rev. Lett. **21**, 1572
Eastman, E.A., J.L. Freeouf, M. Erbudak, 1974: in *Proc. Int. Conf. Tetr. Bonded Amorphous Semic., Yorktown Heights*, ed. by M.H. Brodsky, S. Kirkpatrick, D. Weaire, A.I.P. Conf. Proc. Series 20 (American Institute of Physics, New York) p. 95
Economou, E.N., N. Bacalis, 1987: J. Non-Cryst. Sol. **97/98**, 101
Economou, E.N., M.H. Cohen, 1970: Phys. Rev. Lett. **25**, 1445
Economou, E.N., M.H. Cohen, 1972: Phys. Rev. **B5**, 2931
Economou, E.N., C.M. Soukoulis, M.H. Cohen, S. John, 1987: in *Disordered Semiconductors (Fritzsche Festschrift)* ed. by M.A. Kastner, G.A. Thomas, S.R. Ovshinski (Plenum Press, New York) p. 681
Emin, D., 1973: in *Electronic and Structural Properties of Amorphous Semiconductors*, ed. by P.G. LeComber, J. Mort (Academic Press, London) p. 261
Emin, D., 1974: Phys. Rev. Lett. **32**, 303
Emin, D., 1975: Phys. Rev. Lett. **35**, 882
Emin, D., 1983: Comm. Sol. State Phys. **11**, 35; **11**, 59
Emin, D., T. Holstein, 1976: Phys. Rev. Lett. **36**, 323
Friedman, L., 1971: J. Non-Cryst. Sol. **6**, 329
Fritzsche, H., 1958: J. Phys. Chem. Solids **6**, 69
Fritzsche, H., 1959: Phys. Rev. **115**, 336
Fritzsche, H., 1960: Phys. Rev. **119**, 1899
Fritzsche, H., 1980: Solar Energy Mat. **3**, 447
Fritzsche, H., 1987: J. Non-Cryst. Sol. **97/98**, 95
Fritzsche, H., M. Cuevas, 1960: Phys. Rev. **119**, 1236
Grant, A.J., E.A. Davis, 1974: Solid State Commun. **15**, 563
Greenwood, D.A., 1958: Proc. Phys. Soc. **71**, 585
Grigorovici, R., 1973: in *Electronic and Structural Properties of Amorphous Semiconductors*, ed. by P.G. LeComber, J. Mort (Academic Press, London) p. 191
Grünewald, M., B. Movaghar, B. Pohlmann, D. Würtz, 1985: Phys. Rev. **B32**, 8191
Grünewald, M., P. Thomas, 1979: phys. stat. sol. (b) **94**, 125
Halperin, B.I., M. Lax, 1966: Phys. Rev. **148**, 722
Hamilton, E.M., 1972: Phil. Mag. **26**, 1043
Hauschildt, D., W. Fuhs, H. Mell, 1982: phys. stat. sol. (b) **111**, 171
Henderson, D., F. Hermann, 1972: J. Non-Cryst. Sol. **8–10**, 359

Hermann, F., J.P. Van Dyke, 1968: Phys. Rev. Lett. **21**, 1575
Hindley, N.K., 1970: J. Non-Cryst. Sol. **5**, 17 and 31
Holstein, T., 1959: Ann. Phys. (N.Y.) **8**, 325 and 343
Hubbard, J., 1964: Proc. Royal Soc. **A277**, 237
Hung, C.S., J.R. Gleissman, 1950: Phys. Rev. **79**, 726
Imry, Y., 1980: Phys. Rev. Lett. **44**, 469
Ioffe, A.F., A.R. Regel, 1960: Prog. Semic. **4**, 237
Joannopoulos, J.D., M.H. Cohen, 1973: Phys. Rev. B**7**, 2644; Phys. Rev. B**8**, 2733
Kakalios, J., R.A. Street, 1986: Phys. Rev. B**34**, 6014
Kalbitzer, S., G. Müller, P.G. LeComber, W.E. Spear, 1980: Phil. Mag. **B41**, 439
Kastner, M.A., 1985: J. Non-Cyst. Sol. **77/78**, 1173
Kirkpatrick, S., 1973: Rev. Mod. Phys. **45**, 574
Knights, J.C., 1984: in *Hydrogenated Amorphous Silicon*, ed. by J.D. Joannopoulos, G. Lucovsky, Topics Appl. Phys., Vol. **55** (Springer Verlag, Berlin, Heidelberg, New York) p. 5
Knights, J.C., G. Lucovsky, R.J. Nemanich, 1978: Phil. Mag. **B 37**, 467
Knights, J.C., R.A. Lujan, 1979: Appl. Phys. Lett. **35**, 244
Kramer, B., K. Maschke, P. Thomas, J. Treusch, 1970: Phys. Rev. Lett. **25**, 1020
Kubo, R.J., 1957: J. Phys. Soc. Japan, **12**, 570
Kumeda, M., Y. Yonezawa, K. Nakazawa, S. Ueda, T. Shimizu, 1983: Jpn. J. Appl. Phys. **22**, L 194
Lannin, J.S., 1987: J. Non-Cryst. Sol. **97/98**, 39
Leadbetter, A.J., A.A.M. Rashid, N. Colenutt, A.F. Wright, J.C. Knights, 1981: Solid State Commun. **38**, 957
Leadbetter, A.J., A.A.M. Rashid, R.M. Richardson, A.F. Wright, J.C. Knights, 1980: Solid State Commun. **33**, 973
LeComber, P.G., W.E. Spear, 1970: Phys. Rev. Lett. **25**, 509
LeComber, P.G., A. Madan, W.E.Spear, 1972: J. Non-Cryst. Sol. **11**, 219
Lee, P.A., T.V. Ramakrishnan, 1985: Rev. Mod. Phys. **57**, 287
Ley, L., 1984: in *Semiconductors and Semimetals*, ed. by R.K. Willardson, A.C. Beer, Vol. **21b** (Academic Press, New York) p. 385
Ley, L., S. Kowalczyk, P. Pollak, D.A. Shirley, 1972: Phys. Rev. Lett. **29**, 1088
Liciardello, D.C., D.J. Thouless, 1975: J. Phys. **C8**, 4157
Liciardello, D.C., D.J. Thouless, 1978: J. Phys. **C11**, 925
Lucovsky, G., 1974: in *Proc. 5th Int. Conf. on Physics of Amorphous and Liquid Semic.*, ed. by J. Stuke, W. Brenig (Taylor and Francis, London) p. 1099
Lucovsky, G., T.M. Hayes, 1979: in *Hydrogenated Amorphous Silicon*, ed. by M.H. Brodsky, Topics Appl. Phys., Vol. **36** (Springer, Berlin, Heidelberg, New York) p. 215
MacKinnon, A., B. Kramer, 1981: Phys. Rev. Lett. **47**, 1546
Maschke, K., P. Thomas, 1970: phys. stat. sol. **41**, 743
Maschke, K., H. Overhof, P. Thomas, 1974: in *Tetrahedrally Bonded Amorphous Semiconductors*, ed. by M.A. Brodski, S. Kirkpatrick, D. Weaire, AIP Conference Series 20 (American Institute of Physics, New York) p. 120
Miller, A., E. Abrahams, 1960: Phys. Rev. **120**, 745
Moss, S.C., J.F. Graczyk, 1969: Phys. Rev. Lett. **23**, 1167
Moss, S.C., J.F. Graczyk, 1970: in *Proc. 10th Int. Conf. on the Physics of Semic.*, ed. by S.P. Keller, J.C. Hensel, F. Stern (United States Atomic Energy Agency, Washington D.C.) p. 658
Mott, N.F., 1956: Can. J. Phys. **34**, 1356
Mott, N.F., 1966: Phil. Mag. **13**, 989
Mott, N.F., 1967: Adv. Phys. **16**, 49
Mott, N.F., 1968: J. Non-Cryst. Sol. **1**, 1
Mott, N.F., 1969: Phil. Mag. **19**, 835
Mott, N.F., 1974: *Metal Insulator Transitions* (Taylor and Francis, London)
Mott, N.F., 1987: *Conduction in Non-Crystalline Material* (Clarendon Press, Oxford)
Mott, N.F., 1987: J. Non-Cryst. Sol. **97/98**, 531

Mott, N.F., 1988: *private communication*
Mott, N.F., E.A. Davis, 1971, 1979: *Electronic Processes in Non–Crystalline Materials* (Clarendon Press, Oxford)
Moullet, I., K. Maschke, H. Kunz, 1985: J. Non–Cryst. Sol. **77/78**, 17
Movaghar, B., 1981: J. de Physique (Paris) **42**, C4—73
Movaghar, B., M. Grünewald, B. Pohlmann, D. Würtz, W. Schirmacher, 1983: J. Stat. Phys. **30**, 315
Movaghar, B., W. Schirmacher, 1981: J. Phys. **C14**, 859
Müller, G., S. Kalbitzer, W.E. Spear, P.G. LeComber, 1977: in *Amorphous and Liquid Semiconductors*, ed. by W.E. Spear (CICL, Edinburgh) p. 442
Nagel, S., P. Thomas, 1977: in *Amorphous and Liquid Semiconductors*, ed. by W.E. Spear (CICL, Edinburgh) p. 266
Onsager, L., 1931: Phys. Rev. **37**, 405; Phys. Rev. **38**, 2265
Overhof, H., 1975: phys. stat. sol. (b) **67**, 709
Overhof, H., 1976: in *Festkörperprobleme XVI* , ed. by J. Treusch (Vieweg, Braunschweig) p. 239
Overhof, H., P. Thomas, 1976: in *Electronic Phenomena in Non–Cryst. Semic.*, ed. by B.T. Kolomiets (Nauka, Leningrad) p. 107
Pantelides, S.T., 1986: Phys. Rev. Lett. **57**, 2979
Phillips, J.C., 1979: Phys. Rev. Lett. **42**, 1151
Pierce, D.T., W.E. Spicer, 1972: Phys. Rev. B **5**, 3017
Pike, G.E., C.H. Seager, 1974: Phys. Rev. B **10**, 1421
Polk, D.E., 1971: J. Non–Cryst. Sol. **5**, 365
Polk, D.E., D.S. Boudreaux, 1973: Phys. Rev. Lett. **31**, 92
Pollak, M., 1972: J. Non–Cryst. Sol. **11**, 1
Pollak, M., T.H. Geballe, 1961: Phys. Rev. **122**, 1742
Reimer, J.A., R.W. Vaugham, J.C. Knights, 1980: Phys. Rev. Lett. **44**, 193
Reimer, J.A., R.W. Vaugham, J.C. Knights, 1981: Phys. Rev. **B24**, 3660
Ribbing, C.G., D.T. Pierce, W.E. Spicer, 1971: Phys. Rev. **B4**, 4417
Schiff, E.A., P.D. Persans, H. Fritzsche, V. Akopyan, 1981: Appl. Phys. Lett. **38**, 92
Schönhammer, K., 1971: Phys. Lett. A **36**, 181
Shante, V.K.S., S. Kirkpatrick, 1971: Adv. Physics **20**, 325
Shevshik, N.J., W. Paul, 1972: J. Non–Cryst. Sol. **8–10**, 381
Shklovskii, B.I., A.L. Efros, 1984: *Electronic Properties of Doped Semiconductors*, Springer Ser. Solid–State Sciences, Vol. **45** (Springer, Berlin, Heidelberg, New York)
Singh, J., 1981: Phys. Rev. **B23**, 4156
Smith Jr., J.E., M.H. Brodsky, B.L. Crowder, M.I. Nathan, A. Pinczuk, 1971: Phys. Rev. Lett. **26**, 642
Soukoulis, C.M., M.H. Cohen, E.N. Economou, 1984: Phys. Rev. Lett. **53**, 616
Soukoulis, C.M., M.H. Cohen, E.N. Economou, A.D. Zdetsis, 1985: J. Non–Cryst. Sol. **77/78**, 47
Spear, W.E., P.G. LeComber, 1975: Solid State Comm. **17**, 1193
Spear, W.E., P.G. LeComber, S. Kalbitzer, G. Müller, 1979: Phil. Mag. **B39**, 159
Steinhardt, P., R. Alben, W. Weaire, 1974: J. Non–Cryst. Sol. **15**, 199
Street, R.A., D.K. Biegelsen, J.C. Knights, 1981: Phys. Rev. **B24**, 969
Street, R.A., J. Kakalios, C.C. Tsai, T.M. Hayes, 1987: Phys. Rev. **B35**, 1316
Stuke, J., 1976: in *Electronic Phenomena in Non–Cryst. Semic.*, ed. by B.T. Kolomiets (Nauka, Leningrad) p. 193
Stutzmann, M., R.A. Street, 1985: Phys. Rev. Lett. **54**, 1836
Tauc, J., 1969: in *The Optical Properties of Solids* ed. by F. Abeles (North–Holland, Amsterdam) p. 277
Temkin, R.J., 1978: J. Non–Cryst. Sol. **28**, 23
Temkin, R.J., W. Paul, G.A.N. Connell, 1973: Adv. Phys. **22**, 531
Theye, M.-L., A. Gheorghiou, T. Rappenau, A. Lewis, 1980: J. Physique (Paris) **41**, 1173
Thouless, D.J., 1974: Phys. Rev. **C13**, 94
Thouless, D.J., 1977: Phys. Rev. Lett. **39**, 1167

Thouless, D.J., 1980: J. Non-Cryst. Sol. **35**, 3
Turnbull, D., D.E. Polk, 1972: J. Non-Cryst. Sol. **8–10**, 19
Urbach, F., 1953: Phys. Rev. **92**, 1324
Vollhardt, D., P. Wölfle, 1982: Phys. Rev. Lett. **48**, 699
Wihl, M., M. Cardona, J. Tauc, 1972: J. Non-Cryst. Sol. **8–10**, 172
Wooten, F., D. Weaire, 1987: in *Solid State Physics*, Vol. 40, ed. by H. Ehrenreich, F. Seitz, D. Turnball (Academic Press, New York) p. 1
Würtz, D., P. Thomas, 1978: phys. stat. sol. (b) **88**, K 73
Yamasaki, S., A. Matsuda, K. Tanaka, 1982: Jpn. J. Appl. Phys. **21**, L 789
Yoshino, S., M. Okazaki, 1977: J. Phys. Soc. Japan **43**, 415
Zachariasen, W.H., 1932: J. Am. Chem. Soc. **54**, 3841
Zanzucchi, P.J., 1984: in *Semiconductors and Semimetals*, ed. by R.K. Willardson, A.C. Beer, Vol. **21 b** (Academic Press, New York) p. 113
Ziman, J.M., 1968: J. Phys. **C1**, 1532
Zvyagin, I.P., 1973: phys. stat. sol. (b) **58**, 443

Chapter 3

Anderson, D.A., W. Paul, 1981: Phil. Mag. **B 44**, 187
Anderson, D.A., W. Paul, 1982: Phil. Mag. **B 45**, 1
Ast, D.G., M.H. Brodsky, 1979: in *Proc. 14th Int. Conf. on the Physics of Semic.*, Institute of Physics Conf. Series **43**, ed. by B.L.H. Wilson (Institute of Physics, Bristol) p. 1159
Ast, D.G., M.H. Brodsky, 1980: Phil. Mag. **B 41**, 273
Austin, I.G., N.F. Mott, 1969: Adv. Phys. **18**, 41
Beyer, W., R. Fischer, H. Overhof, 1979: Phil. Mag. **B 39**, 205
Beyer, W., R. Fischer, H. Wagner, 1979: J. Electron. Mat. **8**, 177
Beyer, W., A. Medeišis, H. Mell, 1977: Commun. Phys. **2**, 121
Beyer, W., H. Mell, 1977: in *Amorphous and Liquid Semiconductors*, ed. by W.E. Spear (CICL, Edinburgh) p. 333
Beyer, W., H. Mell, 1981: Solid State Commun. **38**, 891
Beyer, W., H. Mell, H. Overhof, 1977: in *Amorphous and Liquid Semiconductors*, ed. by W.E. Spear (CICL, Edinburgh) p. 328
Beyer, W., H. Mell, H. Overhof, 1981: J. Physique (Paris) **42**, **C4**–103
Beyer, W., H. Overhof, 1979: Solid State Commun. **31**, 1
Beyer, W., H. Overhof, 1984: in *Semiconductors and Semimetals*, Vol. **21c**, ed. by R.K. Willardson, A.C. Beer (Academic Press, New York) p. 257
Beyer, W., B. Stritzker, H. Wagner, 1980: J. Non-Cryst. Sol. **35/36**, 321
Beyer, W., J. Stuke, 1974: in *Proc. 5th Int. Conf. on Physics of Amorphous and Liquid Semic.*, ed. by J. Stuke, W. Brenig (Taylor and Francis, London) p. 251
Beyer, W., J. Stuke, 1975: phys. stat. sol. (b) **30**, 511
Beyer, W., J. Stuke, H. Wagner, 1975: phys. stat. sol. (a) **30**, 231
Beyer, W., H. Wagner, 1981: J. Physique (Paris) **42**, **C4**–783
Beyer, W., H. Wagner, 1982: J. Appl. Phys. **53**, 8745
Den Boer, W., 1981: J. Physique (Paris) **42**, **C4**—451
Bort, M., 1986: Diploma Thesis, Marburg (unpublished)
Butcher, P.N., 1985: in *Amorphous Solids and the Liquid State*, ed. by N. Marsh, R.A. Street, M. Tosi (Plenum Press, New York) p. 311
Carlson, D.E., C.R. Wronski, 1979: in *Amorphous Semiconductors*, ed. by M.H. Brodski, Topics Appl. Phys., Vol. **36** (Springer, Berlin, Heidelberg, New York) Chap. 10
Chittick, R.C., J.H. Alexander, H.F. Sterling, 1969: J. Electrochem. Soc. **116**, 77
Clark, A.H., 1967: Phys. Rev. **154**, 750
Dersch, H., N.M. Amer, 1984: Appl. Phys. Lett. **45**, 272
Djamdji, F., P.G. LeComber, 1987: Phil. Mag. **B 56**, 31
Emin, D., 1977: Phil. Mag. **35**, 1189; in *Proc. 7th Int. Conf. on Amorphous and Liquid Semiconductors*, ed. by W.E. Spear (CICL, Edinburgh) p. 249
Fritzsche, H., 1980: Sol. Energy Mat. **3**, 447

Fuhs, W., 1984: in *Festkörperprobleme XXIV*, ed. by P. Grosse (Vieweg, Braunschweig) p. 133
Ghiassy, F., D.I. Jones, A.D. Stewart, 1985: Phil. Mag. **B 52**, 139
Grigorovici, R., N. Croitorou, A. Dévényi, 1967: phys. stat. sol. **23**, 621
Gross, V., H.W. Grueningen, E. Niemann, R. Fischer, 1987: J. Non-Cryst. Sol. **97/98**, 643
Grünewald, M., B. Movaghar, B. Pohlmann, D. Würtz, 1985: Phys. Rev. **B32**, 8191
Grünewald, M., B. Pohlmann, B. Movaghar, D. Würtz, 1984: Phil. Mag. **B 49**, 341
Grünewald, M., P. Thomas, D. Würtz, 1981: J. Phys. **C14**, 4083
Hauschildt, D., 1982: Ph. D. Thesis, Marburg (unpublished)
Hauschildt, D., W. Fuhs, H. Mell, 1982: phys. stat. sol. (b) **111**, 171
Hauschildt, D., W. Fuhs, H. Mell, K. Weber, 1982: in *Proc. 4th EC Photovoltaic Conference*, ed. by W.H. Bloss, G. Grassi (Reidel, Dordrecht) p. 448
Hauschildt, D., M. Stutzmann, J. Stuke, H. Dersch, 1982: Sol. Energy Mat. **8**, 319
Ioffe, A.F., A.R. Regel, 1960: Prog. Semic. **4**, 237
Irsigler, P., D. Wagner, D.J. Dunstan, 1983: J. Phys. **C16**, 6605
Jan, Z.S., R.H. Bube, J.C. Knights, 1979: J. Electron. Mat. **8**, 47
Jang, J., C. Lee, 1983: J. Non-Cryst. Sol. **59/60**, 281
Jones, D.I., P.G. LeComber, W.E. Spear, 1977: Phil. Mag. **36**, 541
Jones, D.I., W.E. Spear, P.G. LeComber, 1976: J. Non-Cryst. Sol. **20**, 259
Jones, D.I., W.E. Spear, P.G. LeComber, S. Li, R. Martins, 1979: Phil. Mag. **B 39**, 147
Kagawa, T., Y. Muramatsu, 1986: J. Non-Cryst. Sol. **81**, 261
Kakalios, J., R.A. Street, 1986: Phys. Rev. **B34**, 6014
Krühler, W., H. Pfleiderer, R. Plättner, W. Stetter, 1984: AIP Conf. Ser., Vol. 120, ed. by P.C. Taylor, S.G. Bishop (American Institute of Physics, New York) p. 311
Lampert, M.A., P. Mark, 1970: *Current Injection in Solids* (Academic Press, New York)
LeComber, P.G., D.I. Jones, W.E. Spear, 1977: Phil. Mag. **36**, 1173
LeComber, P.G., W.E. Spear, 1970: Phys. Rev. Lett. **25**, 509
Mackenzie, K.D., P.G. LeComber, W.E. Spear, 1982: Phil. Mag. **B 46**, 377
Marshall, J.M., 1977: Phil. Mag. **35**, 959
Marshall, J.M., 1983: Rep. Prog. Phys. **46**, 1235
Marshall, J.M., R.A. Street, M.J. Thompson, 1986: Phil. Mag. **B 54**, 51
Marshall, J.M., R.A. Street, M.J. Thompson, W.B. Jackson, 1988: Phil. Mag. **B 57**, 387
Meyer, W., H. Neldel, 1937: Z. Techn. Phys. **12**, 588
Michiel, H., J.M. Marshall, G.J. Adriaennsens, 1983: Phil. Mag. **B 48**, 187
Monroe, D., 1985: Phys. Rev. Lett. **54**, 146
Movaghar, B., 1981: J. Physique (Paris) **42**, **C4** – 73
Movaghar, B., B. Pohlmann, G.W. Sauer, 1980: phys. stat. sol. (b) **97**, 533
Movaghar, B., B. Pohlmann, W. Schirmacher, 1980: Phil. Mag. **B 41**, 49
Movaghar, B., L. Schweitzer, 1978: J. Phys. **C 11**, 125
Müller, G., F. Demond, S. Kalbitzer, H. Demjantschitsch, H. Mannsberger, W.E. Spear, P.G. LeComber, R.A. Gibson, 1980: Phil. Mag. **B 41**, 571
Noolandi, J., 1977: Phys. Rev. **B16**, 4466
Orenstein, J., M. Kastner, 1981: Phys. Rev. Lett. **46**, 1421
Orton, J.W., M.J. Powell, 1984: Phil. Mag. **B 50**, 11
Ortuno, M., M. Pollak, 1983: J. Non-Cryst. Sol. **59/60**, 53
Overhof, H., 1976: in *Festkörperprobleme XVI*, ed. by J. Treusch (Vieweg, Braunschweig), p. 239
Overhof, H., W. Beyer, 1983: Phil. Mag. **B 47**, 377
Overhof, H., W. Beyer, 1984: Phil. Mag. **B 49**, L 9
Overhof, H., P. Thomas, 1976: in *Electronic Phenomena in Non-Crystalline Semiconductors*, ed. by B.T. Kolomiets (Nauka, Leningrad) p. 107
Paul, W., D.A. Anderson, 1981: Solar Energy Mat. **5**, 229
Pollak, M., 1971: Phil. Mag. **23**, 519
Pollak, M., 1977: Phil. Mag. **36**, 1157; in *Amorphous and Liquid Semiconductors*, ed. by W.E. Spear (CICL, Edinburgh) p. 219
Pollak, M., T.H. Geballe, 1961: Phys. Rev. **122**, 1742

Rehm, W., R. Fischer, J. Stuke, H. Wagner, 1977: phys. stat. sol. (b) **79**, 539
Roilos, M., 1978: Phil. Mag. **B 38**, 477
Scharfe, M.E., 1970: Phys. Rev. B**2**, 5025
Schauer, F., O. Smeškal, 1987: J. Non-Cryst. Sol. **97/98**, 748
Scher, H., E.W. Montroll, 1975: Phys. Rev. B**12**, 2455
Schmidlin, F.W., 1977: Phys. Rev. B**16**, 2362
Schropp, R.E.I., J. Snijder, J.F. Verwey, 1986: J. Appl. Phys. **60**, 643
Schropp, R.E.I., G.J.M. Brouwer, J.F. Verwey, 1985: J. Non-Cryst. Sol. **77/78**, 511
Schumacher, R., P. Thomas, K. Weber, W. Fuhs, 1987: Solid State Commun. **62**, 15
Schumacher, R., P. Thomas, K. Weber, W. Fuhs, F. Djamdi, P.G. LeComber, R.E.I. Schropp, 1988: Phil. Mag. **B 58**, 389
Shimakawa, K., A.R. Long, M.J. Anderson, O. Imagawa, 1987: J. Non-Cryst. Sol. **97/98**, 623
Shimakawa, A.R. Long, O. Imagawa, 1987: Phil. Mag. **B 56**, L79
Silver, M., L. Cohen, 1977: Phys. Rev. B**15**, 3276
Silver, M., L. Cohen, D. Adler, 1981: Phys. Rev. B**24**, 4855
Silver, M., N.C. Giles, E. Snow, 1982: Solar Energy Mat. **8**, 303
Silver, M., E. Snow, N.C. Giles, M.P. Shaw, V. Canella, S. Paysan, R. Ross, S. Hudgens, 1983: Physica **B 117/118**, 905
Silver, M., E. Snow, B. Wright, L. Moore, V. Canella, R. Ross, S. Paysan, M.P. Shaw, D. Adler, 1983: Phil. Mag. **B 47**, L 39
Šmíd, V., J. Mareš, N.M. Dung, L. Štourač, J. Křistofik, 1985: J. Non-Cryst. Sol. **77/78**, 311
Solomon, I., 1981: in *Fundamental Physics of Amorphous Semiconductors*, ed. by F. Yonezawa, Springer Series in Solid State Sciences, Vol. **25** (Springer, Berlin, Heidelberg, New York) p. 33
Solomon, I., T. Dietl, D. Kaplan, 1978: J. Physique (Paris), **39**, 1241
Solomon, I., R. Benferhat, H. Tran-Quoc, 1984: Phys. Rev. B**30**, 3422
Spear, W.E., 1957: Proc. Phys. Soc. London **B 70**, 669
Spear, W.E., 1969: J. Non-Cryst. Sol. **1**, 197
Spear, W.E., D. Allen, P.G. LeComber, A. Ghaith, 1980: Phil. Mag. **B 41**, 419
Spear, W.E., P.G. LeComber, 1972: J. Non-Cryst. Sol. **8 – 10**, 727
Spear, W.E., P.G. LeComber, 1975: Solid State Commun. **17**, 1193
Spear, W.E., H. Steemers, 1983: Phil. Mag. **B 47**, L 77 and L 107
Staebler, D.L., C.R. Wronski, 1977: Appl. Phys. Lett. **31**, 292
Staebler, D.L., C.R. Wronski, 1980: J. Appl. Phys. **51**, 3262
Sterling, H.F., R.C.G. Swann, 1965: Sol. State Electr. **8**, 653
Street, R.A., J. Kakalios, T.M. Hayes, 1986: Phys. Rev. B**34**, 3030
Street, R.A., J. Kakalios, M. Hack, 1988: Phys. Rev. B**38**, 5608
Street, R.A., J. Kakalios, C.C. Tsai, T.M. Hayes, 1987: Phys. Rev. B**35**, 1316
Stuke, J., 1976: in *Electronic Phenomena in Non-Crystalline Semiconductors*, ed. by B.T. Kolomiets (Nauka, Leningrad) p. 193
Tanielian, M., H. Fritzsche, C.C. Tsai, E. Symbalisty, 1978: Appl. Phys. Lett. **33**, 353
Tiedje, T., 1984: in *Hydrogenated Amorphous Silicon II*, ed. by J.D. Joannopoulos, G. Lucovsky, Topics in Applied Physics, Vol. **56** (Springer, Berlin, Heidelberg, New York) p. 207
Tiedje, T., 1984: in *Semiconductors and Semimetals*, Vol. **21c**, ed. by R.K. Willardson, A.C. Beer (Academic Press, New York) p. 207
Tiedje, T., A. Rose, 1980: Solid State Commun. **37**, 49
Wagner, D., P. Irsigler, D.J. Dunstan, 1983: J. Non-Cryst. Sol. **59/60**, 413
Walley, P.A., 1968: Thin Solid Films **2**, 327; Ph. D. Thesis, London (unpublished)
Walley, P.A., A.K. Jonscher, 1968: Thin Solid Films **1**, 367
Weller, D., H. Mell, L. Schweitzer, J. Stuke, 1981: J. Physique (Paris), **42, C4 – 143**

Chapter 4

Anderson, P.W., 1972: Nature Phys. Sci. (London) **235**, 163
Aoki, H., 1982: Physica **114A**, 538

Aoki, H., 1983: J. Phys. **C16**, L 205
Belitz, D., A. Gold, W. Götze, 1981: Z. Phys. B**44**, 273
Belitz, D., W. Götze, 1982: J. Phys. **C15**, 981
Belitz, D., W. Götze,1983: Phys. Rev. B**28**, 5445
Brenig, W., G.H. Döhler, P. Wölfle, 1971: Z. Phys. **246**, 1
Brenig, W., G.H. Döhler, P. Wölfle, 1973: Z. Phys. **258**, 381
Cohen, M.H., E.N. Economou, C.M. Soukoulis, 1983: J. Non-Cryst. Sol. **59/60**, 15; Phys. Rev. Lett. **51**, 1202
Cohen, M.H., E.N. Economou, C.M. Soukoulis, 1984: Phys. Rev. B**29**, 4496 and 4500
Dersch, U., 1986: Ph. D. Thesis, Marburg (unpublished)
Dersch, U., P. Thomas, 1985: J. Phys. **C18**, 5815
Economou, E.N., C.M. Soukoulis, M.H. Cohen, S. John, 1987: in *Disordered Semiconductors*, ed. by M.A. Kastner, G.A. Thomas, S.R. Ovshinsky (Plenum, New York) p. 681
Emin, D., T. Holstein, 1976: Phys. Rev. Lett. **36**, 323
Fenz, P., H. Müller, H. Overhof, P. Thomas, 1985: J. Phys. **C18**, 3191
Girvin, S.M., M. Jonson, 1980: Phys. Rev. B**22**, 3583
Götze, W., 1978: Solid State Commun. **27**, 1393
Götze, W., 1979: J. Phys. **C12**, 1279
Götze. W., 1981: Phil. Mag. **43**, 219
Götze, W., 1985: in *Localization, Interaction, and Transport Phenomena*, Solid-State Sci., Vol. **61**, ed. by B. Kramer, G. Bergmann, Y. Bruynseraede (Springer, Berlin, Heidelberg, New York) p. 62
Hall, G.G., 1952: Phil. Mag. **43**, 338
Hall, G.G., 1958: Phil. Mag. **3**, 429
Imry, Y., 1980: Phys. Rev. Lett. **44**, 469
Jonson, M., S.M. Girvin, 1979: Phys. Rev. Lett. **43**, 1447
Kikuchi, M., 1983: J. Non-Cryst. Sol. **59/60**, 25
McInnes, J.A., P.N. Butcher, J.D. Clark, 1980: Phil. Mag. **B41** , 1
Miller, A., E. Abrahams, 1960: Phys. Rev. **120**, 745
Mott, N.F., 1972: Phil. Mag. **26**, 1015
Movaghar, B., B. Pohlmann, G.W. Sauer, 1980: phys. stat. sol. (b)**97**, 533
Movaghar, B., B. Pohlmann, W. Schirmacher, 1980: Phil. Mag. **B41**, 49
Movaghar, B., B. Pohlmann, D. Würtz, 1981: J. Phys. **C14**, 5127
Movaghar, B., W. Schirmacher, 1981: J. Phys. **C14**, 859
Müller, H., P. Thomas, 1983: Phys. Rev. Lett. **51**, 702
Müller, H., P. Thomas, 1984: J. Phys. **C17**, 5337
Soukoulis, C.M., M.H. Cohen, E.N. Economou, 1984: Phys. Rev. Lett. **53**, 616
Soukoulis, C.M., M.H. Cohen, E.N. Economou, A. Zdetsis, 1985: J. Non-Cryst. Sol. **77/78**, 47
Soukoulis, C.M., E.N. Economou, 1984: Phys. Rev. Lett. **52**, 565
Thouless, D.J., 1977: Phys. Rev. Lett. **39**, 1167
Viščor, P., 1983: Phys. Rev. B**28**, 927
Weaire, D., M.F. Thorpe, 1971: Phys. Rev. B4, 2508 and 3518

Chapter 5

Belitz, D., W. Götze, 1983: Phys. Rev. B**28**, 5445
Belitz, D., W. Schirmacher, 1983: J. Phys. **C16**, 913
Dersch, U., B. Pohlmann, P. Thomas, 1983: J. Phys. **C16**, 3725
Economou, E.N., 1983: *Green's Functions in Quantum Physics*, Springer Ser. Solid-State Sci., Vol. **7** (Springer, Berlin, Heidelberg, New York)
Economou, E.N., C.M. Soukoulis, 1983: Phys. Rev. B**28**, 1093
Economou, E.N., C.M. Soukoulis, M.H. Cohen, A.D. Zdetsis, 1985: Phys. Rev. **B31**, 6172
Economou, E.N., C.M. Soukoulis, A.D. Zdetsis, 1984: Phys. Rev. B**30**, 1686
Economou, E.N., A.D. Zdetsis, D.A. Papaconstantopoulos, 1985: J. Non-Cryst. Sol. **77/78**, 147

Emin, D., 1974: Phys. Rev. Lett. **32**, 303
Forster, D., 1975: *Hydrodynamic Fluctuations, Broken Symmetry and Correlation Functions* (Benjamin, Reading M.A.)
Girvin, S.M., M. Jonson, 1980: Phys. Rev. B**22**, 3583
Götze, W., 1978: Solid State Commun. **27**, 1393
Götze, W., 1979: J. Phys. **C12**, 1279
Götze, W., 1981: Phil. Mag. **B 43**, 219
Haken, H., G. Strobel, 1967: *Exact Treatment of Coherent and Incoherent Triplet Exciton Migration*, in "The Triplet State", ed. by A. Zahlan (Cambridge Univ. Press, London) p. 311
Holstein, T., 1959: Ann. Phys. **N.Y. 8**, 325 and 343
Jonson, M. S.M. Girvin, 1979: Phys. Rev. Lett. **43**, 1447
Kadanoff, L.P., G. Baym, 1962: *Quantum Statistical Mechanics* (Benjamin, New York)
Kawasaki, K., 1966: Phys. Rev. **150**, 291
Kenkre, V.M., P. Reinecker, 1982: *Exciton Dynamics in Molecular Crystals and Aggregates*, Springer Tracts Mod. Phys., Vol. **94** (Springer, Heidelberg, Berlin, New York)
Loring, R.F., S. Mukamel, 1986: J. Chem. Phys. **85**, 1950
Mahan, G.D., 1983: *Many-Particle Physics* (Plenum, New York)
Miller, A., E. Abrahams, 1960: Phys. Rev. **120**, 745
Mori, H., 1965: Progr. Theor. Phys. **33**, 423
Mott, N.F., 1972: Phil. Mag. **26**,1015
Movaghar, B., B. Pohlmann, G.W. Sauer, 1980: phys. stat. sol. (b) **97**, 533
Movaghar, B., B. Pohlmann, W. Schirmacher, 1980: Phil. Mag. **B41**, 49
Movaghar, B., B. Pohlmann, D. Würtz, 1981: J. Phys. **C14**, 5127
Movaghar, B., W. Schirmacher, 1981: J. Phys. **C14**, 859
Müller, H., P. Thomas, 1984: J. Phys. **C17**, 5337
Papaconstantopoulos, D.A., E.N. Economou, 1980: Phys. Rev. B**22**, 2903
Papaconstantopoulos, D.A., E.N. Economou, 1981: Phys. Rev. B**24**, 7233
Schnakenberg, J., 1966: Z. Phys. **190**, 209
Schnakenberg, J., 1968: phys. stat. sol. **28**, 623
Thomas, P., A. Weller, 1987: J. Non-Cryst. Sol. **97/98**, 245
Thomas, P., A. Weller, 1989: to be published
Vollhardt, D., P. Wölfle, 1980: Phys. Rev. Lett. **45**, 842; Phys. Rev. B**22**, 4666
Wölfle, P., D. Vollhardt, 1982: in *Anderson Localization*, Solid-State Sci., Vol. **39**, ed. by Y. Nagaoka, H. Fukuyama (Springer, Berlin, Heidelberg, New York) p. 26
Zdetsis, A.D., 1986: in *Hydrogen in Disordered and Amorphous Solids*, ed. by G. Bambakidis, R.C. Bowman Jr. (Plenum, New York) p. 27
Zdetsis, A.D., 1987: J. Non-Cryst. Sol. **97/98**, 515
Zdetsis, A.D., E.N. Economou, D.A. Papaconstantopoulos, N. Flytzanis, 1985a: Phys. Rev. B**31**, 2410
Zdetsis, A.D., C.M. Soukoulis, E.N. Economou, G.S. Grest, 1985b: Phys. Rev. B**32**, 7811
Zwanzig, R., 1960: J. Chem. Phys. **33**, 1338
Zwanzig, R., 1961: Phys. Rev. **124**, 983

Chapter 6

Adler, J., M. Silver, 1982: Phil. Mag. **B45**, 307
Beyer, W., H. Overhof, 1984: in *Semiconductors and Semimetals*, Vol. **21c** ed. by R.K. Willardson, A.C. Beer (Academic Press, New York) p. 257
Bonch-Bruevich, V.L., 1962: Fiz. Tverd. Tela **4**, 2660 (English transl.: Sov. Phys. – Solid State **4**, 1953 (1963))
Bonch-Bruevich, V.L., 1970: phys. stat. sol. **42**, 35
Bonch-Bruevich, V.L., 1983: Usp. Fiz. Nauk **104**, 583 (English transl.: Sov. Phys. Usp. **26**, 664)
Fritzsche, H., 1971: J. Non-Cryst. Sol. **6**, 49

Fritzsche, H., 1982: Thin Solid Films **90**, 119
Jäckle, J., 1980: Phil. Mag. **B 41**, 681
Kane, E.O., 1963: Phys. Rev. **131**, 79
Keldysh, L.V., G.P. Proshko, 1963: Fiz. Tverd. Tela **5**, 3378 (English transl.: Sov. Phys. – Solid State **5**, 2481 (1964))
King, H., B. Kramer, A. MacKinnon, 1983: Solid State Commun. **47**, 683
Kirchhoff, L., 1845: Annalen der Physik und Chemie **64**, 497
Kirchhoff, L., 1947: Annalen der Physik und Chemie **72**, 497
Kirkpatrick, S., 1971: Phys. Rev. Lett. **27**, 1722
Knights, J., 1980: J. Non-Cryst. Sol. **35/36**, 159
Knights, J., R.A. Lujan, 1979: Appl. Phys. Lett. **35**, 244
Kramer, B., H. King, A. MacKinnon, 1983: J. Non-Cryst. Sol. **59/60**, 73
Last, B.J., D.J. Thouless, 1971: Phys. Rev. Lett. **27**, 1719
Ley, L., J. Reichardt, R.J. Johnson, 1982: Phys. Rev. Lett. **49**, 1664
Overhof, H., 1981: Phil. Mag. **B 44**, 317
Overhof, H., 1983: J. Non-Cryst. Sol. **59/60**, 57
Overhof, H., 1985: J. Non-Cryst. Sol. **77/78**, 143
Overhof, H., W. Beyer, 1981: Phil. Mag. **B 43**, 433
Seynhaeve, G., G.J. Adriaenssens, H. Michiel, H. Overhof, 1988: Phil. Mag. **B 58**, 421
Shklovskii, B.I., A.L. Efros, 1970: Fiz. Tekh. Poluprov. **4**, 305 (English transl. in Sov. Physics – Semic. **4**, 247)
Shklovskii, B.I., A.L. Efros, 1971: Z. Exp. Theor. Fiz. **60**, 867 (English transl. in Sov. Physics – JEPT **33**, 468)
Shklovskii, B.I., A.L. Efros, 1972: Z. Exp. Theor. Fiz. **62**, 1156 (English transl. in Sov. Physics – JEPT **35**, 610)
Shklovskii, B.I., A.L. Efros, 1984: *Electronic Properties of Doped Semiconductors* (Springer, Berlin, Heidelberg, New York)
Stinchcombe, R.B., 1973: J. Phys. **C6**, L1
Tauc, J., 1970: Mat. Res. Bull. **5**, 721

Chapter 7

Adler, D., 1981: J. de Physique (Paris) **42**, **C4** – 1
Adler, D., 1987: J. Non-Cryst. Sol. **90**, 77
Anderson, P.W., 1975: Phys. Rev. Lett. **34**, 953
Anderson, P.W., 1976: J. Physique (Paris) **C–4**, 339
Beyer, W., H. Mell, H. Overhof, 1977: in *Amorphous and Liquid Semiconductors*, ed. by W.E. Spear (CICL, Edinburgh) p. 328
Beyer, W., H. Overhof, 1979: Solid State Commun. **31**, 1
Busch, G., H. Labhart, 1986: Helv. Phys. Acta **19**, 463
Butcher, P.N., 1984: Phil. Mag. **B 50**, L5
Butcher, P.N., L. Friedmann, 1977: J. Phys. **C 10**, 3803
Drüsedau, T., R. Bindemann, 1986: phys. stat. sol. (b) **136**, K 61
Drüsedau, T., D. Wagner, R. Bindemann, 1987: phys. stat. sol. (b) **140**, K 27
Emin, D., 1977: Solid State Commun. **22**, 409
Emin, D., 1985: Phil. Mag. **B 51**, L 53
Freyland, W., H.P. Pfeifer, F. Hensel, 1974: in *Proc. 5th Int. Conf. on Physics of Amorphous and Liquid Semic.*, ed. by J. Stuke, W. Brenig (Taylor and Francis, London) p. 1327
Fritzsche, H., 1971: J. Non-Cryst. Solids **6**, 49
Fritzsche, H., 1980: Solar Energy Mat. **3**, 447
Grünewald, M., P. Thomas, D. Würtz, 1980: phys. stat. sol. (b) **100**, K 139
Hubbard, J., 1964: Proc. Royal Soc. **A 277**, 237
Irsigler, P., D. Wagner, D.J. Dunstan, 1983: J. Phys. **C16**, 6605
Jackson, W.B., 1988: Phys. Rev. B38, 3595
Jones, D.I., P.G. LeComber, W.E. Spear, 1977: Phil. Mag. **36**, 541

Liciardello, D.C., D.L. Stein, F.D.M. Haldane, 1981: Phil. Mag. **B 43**, 189
Liciardello, D.C., 1981: J. Phys. **C14**, L627
Madelung, O., 1957: in *Handbuch der Physik XX*, ed. by S. Flügge (Springer, Berlin, Heidelberg) p. 1
Mott, N.F., E.A. Davis, 1971, 1979: *Electronic Processes in Non-Crystalline Materials* (Clarendon Press, Oxford)
Mott, N.F., E.A. Davis, R.A. Street, 1975: Phil. Mag. **32**, 961
Narasimhan, K.L., B.M. Arora, 1985: Sol. State Commun. **55**, 615
Overhof, H., W. Beyer, 1981: Phil. Mag. **B 43**, 433
Overhof, H., W. Beyer, 1983: Phil. Mag. **B 47**, 377
Schmutzler, R.W., F. Hensel, 1972: J. Non-Cryst. Sol. **8 – 10**, 718
Schumacher, R., P. Thomas, K. Weber, W. Fuhs, F. Djamdi, P.G. LeComber, R.E.I. Schropp, 1988: Phil. Mag. **B 58**, 389
Spear, W.E., P.G. LeComber, 1976: Phil. Mag. **33**, 935
Street, R.A., N.F. Mott, 1975: Phys. Rev. Lett. **35**, 1293
Watkins, G.D., J.R. Troxel, 1980: Phys. Rev. Lett. **44**, 593

Chapter 8

Adler, R., D. Janes, B.I. Hunsinger, S. Datta, 1981: Appl. Phys. Lett. **38**, 102
Beyer, W., H. Mell, H. Overhof, 1981: J. de Physique (Paris) **42**, **C4** – 103
Beyer, W., J. Stuke, 1974: in *Proc. 5th Int. Conf. on Physics of Amorphous and Liquid Semiconductors*, ed. by J. Stuke, W. Brenig (Taylor and Francis, London) p. 251
Chen, K.J., H. Fritzsche, 1983: J. Non-Cryst. Sol. **58/60**, 444
Cohen, J.D., 1984: in *Semiconductors and Semimetals*, Vol. **21c**, ed. by R.K. Willardson, A.C. Beer (Academic Press, New York) p. 9
Cohen, J.D., D.V. Lang, J.P. Harbison, 1980: Phys. Rev. Lett. **45**, 197
Dersch, H., 1983: Ph. D. Thesis, Marburg (unpublished)
Dersch, H., J. Stuke, J. Beichler, 1981: Appl. Phys. Lett. **38**, 456; phys. stat. sol. (b) **105**, 1265
Foller, M., W. Beyer, J. Herion, H. Wagner, 1986: Surf. Science **178**, 47
Foller, M., J. Herion, W. Beyer, H. Wagner, 1985: J. Non-Cryst. Sol. **77/78**, 979
Fritzsche, H., 1984: Phys. Rev. B**29**, 6672
Fritzsche, H., K.J. Chen, 1983: Phys. Rev. B**28**, 4900
Ghiassy, F., D.I. Jones, A.D. Steward, 1985: Phil. Mag. **B 52**, 139
Gruntz, K., 1981: Ph. D. Thesis, Stuttgart (unpublished)
Hauschildt, D., 1982: Ph. D. Thesis, Marburg (unpublished)
Hauschildt, D., W. Fuhs, H. Mell, K. Weber, 1982: in *Proc. 4th EC Photovoltaic Conference*, ed. by W.H. Bloss, G. Grassi (Reidel, Dordrecht) p. 448
Hauschildt, D., M. Stutzmann, J. Stuke, H. Dersch, 1982: Sol. Energy Mat. **8**, 319
Hoheisel, M., W. Fuhs, 1988: Phil. Mag. **B 57**, 411
Irsigler, P., D. Wagner, D.J. Dunstan, 1983: J. Phys. **C16**, 6605
Jackson, W.B., N.M. Amer, 1982: Phys. Rev. B**25**, 5559
Jackson, W.B., C.C. Tsai, S.M. Kelso, 1985: J. Non-Cryst. Sol. **77/78**, 281
Jackson, W.B., J. Kakalios, 1988: Phys. Rev. B**37**, 1020
Johnson, E.O., 1958: Phys. Rev. **111**, 153
Jones, D.I., P.G. LeComber, W.E. Spear, 1977: Phil. Mag. **36**, 541
Kakalios, J., R.A. Street, 1986: Phys. Rev. B**34**, 6014
Kelvin, Lord, 1898: Phil. Mag. **5**, 45
Kočka, J., M. Vanecek, F. Schauer, 1987: J. Non-Cryst. Sol. **97/98**, 715
Lang, D.V., J.D. Cohen, J.P. Harbison, 1982: Phys. Rev. B**25**, 5285
Ley, L., 1984: in *Semiconductors and Semimetals*, Vol. **21 c**, ed. by R.K. Willardson, A.C. Beer (Academic Press, New York) p.385
Madan, A., P.G. LeComber, W.E. Spear, 1976: J. Non-Crst. Sol. **20**, 239
Marschall, J.M., R.A. Street, M.J. Thompson, 1984: Phys. Rev. B**29**, 2331; J. Non-Cryst. Sol. **66**, 175

Meaudre, R., M. Meaudre, P. Jensen, G. Guirard, 1988: Phil. Mag. Lett. **6**, 315
Mytilineou, E., E.A. Davis, 1977: in *Amorphous and Liquid Semiconductors*, ed. by W.E. Spear (CICL, Edinburgh) p. 632
Overhof, H., 1985: in *Tetrahedrally Bonded Amorphous Semiconductors (Mott Festschrift)*, ed. by H. Fritzsche, D. Adler (Plenum Press, New York) p. 287
Overhof, H., 1987: J. Non–Cryst. Sol. **97/98**, 539
Overhof, H., 1987: in *Disordered Semiconductors (Fritzsche Festschrift)*, ed. by M.A. Kastner, G.A. Thomas (Plenum Press, New York) p. 713
Overhof, H., M. Silver, 1989: to be published
Overhof, H., W. Beyer, 1979: J. Non–Cryst. Sol. **35/36**, 375
Pierz, K., B. Hilgenberg, H. Mell, G. Weiser, 1987: J. Non-Cryst. Sol. **97/98**, 63
Siefert, J.M., G. de Rosny, 1985: J. Non–Cryst. Sol. **77/78**, 531
Spear, W.E., C. Cloude, D. Goldie, P. LeComber, 1987: J. Non-Cryst. Sol. **97/98**, 15
Spear, W.E., P.G. LeComber, 1972: J. Non–Cryst. Sol. **8 – 10**, 727
Staebler, D.L., C.R. Wronski, 1977: Appl. Phys. Lett. **31**, 292
Street, R.A., 1982: Phys. Rev. Lett. **49**, 1187
Street, R.A., 1985: J. Non–Cryst. Sol. **77/78**, 1
Street, R.A., D.K. Biegelsen, 1980: Solid State Commun. **33**, 1151
Street, R.A., J. Kakalios, M. Hack, 1988: Phys. Rev. **B38**, 5608
Street, R.A., J. Kakalios, T.M. Hayes, 1986: Phys. Rev. **B34**, 3030
Street, R.A., J. Kakalios, C.C. Tsai, T.M. Hayes, 1987: Phys. Rev. **B35**, 1316
Street, R.A., J. Zesch, 1984: Phil. Mag. **B50**, L21
Stuke, J., 1984: in *Poly – Microcrystalline and Amorphous Semiconductors*, ed. by P. Pinard, S. Kalbitzer (Editions de Physique, Straßbourg) p. 415
Stutzmann, M., 1986: Phil. Mag. **B 53**, L 15
Stutzmann, M., J. Stuke, 1983: Solid State Commun. **47**, 635
Stutzmann, M., J. Stuke, H. Dersch, 1983: phys. stat. sol. (b) **115**, 141
Takada, J., H. Fritzsche, 1987: Phys. Rev. **B36**, 1706 and 1710
Tanaka, K., S. Yamasaki, 1982: Solar Energy Mat. **8**, 277
Weber, K., M. Grünewald, W. Fuhs, P. Thomas, 1982: phys. stat. sol. (b) **110**, 133
Winer, K., I. Hirabayashi, L. Ley, 1988: Phys. Rev. Lett. **60**, 2697

Author Index*

* This index makes only reference to the first authors of the cited contributions.

Subject Index

Springer Tracts in Modern Physics

Within this long-established series there are several volumes on themes which are related to the subject of this volume and which may be of interest to you and your colleagues:

Volumes 51, 56, 74, 76, 77, 78, 81, 87, 88, 94, 98, 99, 110.

The subject dealt with in this volume is particularly closely related to the themes of the following volumes:

Volume 94

V. M. Kenkre, P. Reinecker

Exciton Dynamics

1982. 226 pages. Hard cover. ISBN 3-540-11318-5

"... an important and useful introduction for theorists and experimentalists alike.
... I regard this as a well-written useful book in the area of theoretical-condensed matter science."

Applied Optics

Volume 95

H. Grabert

Projection Operator Techniques in Nonequilibrium Statistical Mechanics

1982. 164 pages. Hard cover. ISBN 3-540-11635-4

"... an excellent guide book for all physicists applying the powerful techniques of projection operators for solving various particular problems. Though compact, it is easily readible, and so may also serve as an introduction for nonexperts."

Journal de Physique

Springer-Verlag Berlin
Heidelberg New York London
Paris Tokyo Hong Kong

Springer Tracts in Modern Physics

Listed below are some titles that have been commissioned for this series. At the time when this book was made the authors of some of the following books were still busy writing their manuscripts, and in other cases the completed typescripts were still in the process of being transformed into finished books. The titles are listed alphabetically by author:

Albers et al.: **Hydrogen in Materials**
ISBN 3-540-50675-6

V. Devanathan, A. Nagl, H. Überall:
Nuclear Pion Photoproduction
ISBN 3-540-50671-3

A. Hasegawa: **Optical Solitons in Fibers**
ISBN 3-540-50668-3

S. Hunklinger, M. Klinger, K. Shvarts:
Physics of Nonmetallic Glasses
ISBN 3-540-50672-1

P. Mulser: **High Power Laser-Matter Interaction**
ISBN 3-540-50669-1

V. N. Oraevsky et al.: **Artificial Plasma Clouds**
ISBN 3-540-50670-5

Volume 115

B. Poelsema, G. Comsa

Scattering of Thermal Energy Atoms from Disordered Surfaces

1989. 74 figures. Approx. 150 pages. Hard cover. ISBN 3-540-50358-7

A variety of novel applications for the investigation of disordered surfaces by beams of thermal energy atoms are discussed and illustrated by numerous examples. A straightforward semiclassical approach is introduced to yield a remarkably detailed insight into the lateral distributions of diffusive scatterers such as adsorbates, vacancies, and atomic steps.

Springer-Verlag Berlin Heidelberg New York London Paris Tokyo Hong Kong